Praise for **HACKING DARWIN**

"Jamie Metzl's book—ranging through genetics, human destiny, and the future—is an incredible trove of information for scientists, futurists, technologists, entrepreneurs, and virtually every intrigued, interested reader. A gifted and thoughtful writer, Metzl brings us to the frontiers of biology and technology and reveals a world full of promise and peril. This is a must-read book."

—Siddhartha Mukherjee, MD, *New York Times* bestselling author of
The Emperor of All Maladies and *The Gene: An Intimate History*

"In many ways, Jamie Metzl has been preparing *Hacking Darwin* for the last twenty years, and his diligence shines through. Jamie is a gifted writer, and his explanation of genetics is crisp, accurate, and wonderfully engaging. If you can only read one book on the future of our species, this is it."

—Sanjay Gupta MD, bestselling author, neurosurgeon, and
Emmy-award-winning chief medical correspondent (CNN)

"Whether you already run the whitewater of transhumanism or oppose changing one atom of any species, you will see we have a revolution in our midst—one of engineering genes, possibly surpassing the space, atomic, and electronic revolutions in its significance. It is a transformation we ignore at great risk. To help prepare us for what's coming, Jamie Metzl guides us brilliantly down the twisting and ever-changing river of our genetic future in this important and thought-provoking book. Beyond the issues of efficacy and safety, he thoughtfully explores the ultimately more critical issues of diversity, equality, and respect for each other and our common humanity."

—George Church, professor of genetics at Harvard
Medical School and author of *Regenesis*

"Jamie Metzl possesses a nearly superhuman ability to take in vast amounts of research and synthesize it into disruptive, beautifully wrought prose. To borrow a term from the book, *Hacking Darwin* is an intellectual masturbatorium—at once stimulating and consummately satisfying."

—Dan Buettner, National Geographic fellow and *New York Times*
bestselling author of *Blue Zones Solution* and *Blue Zones of Happiness*

"Jamie Metzl's *Hacking Darwin* is an outstanding guide to the most important conversation of our lives—how we humans will hijack our evolutionary process and transcend the limits of our own biology."

—Ray Kurzweil, inventor, futurist, and author of
The Singularity Is Near and *How to Create a Mind*

"Breathtaking advances in artificial intelligence, genomics, and gene editing are radically transforming our understanding of the human body and changing our practice of medicine. But as Jamie Metzl articulates in *Hacking Darwin*, the implications of the genetics revolution extend far beyond health care and into the realm of our very identity as human beings. Metzl lays out the underlying science and thoroughly examines the radical implications and thorny ethical issues raised by these technologies. A genetically altered future will be arriving far sooner and with far greater implications than most people appreciate. *Hacking Darwin* comes at a critically important time, with many scientific developments arising in the field. It is an essential guide as society navigates these developments and a clarion call for the inclusive global dialogue on the future of human genetic engineering we so desperately need."

—Victor Dzau, MD, president of United States
National Academy of Medicine

"Humans taking control of human evolution is happening now and it could be the most significant thing we have done since we learned how to make and use fire. When history looks back on this period, it will not focus on the shiny objects that grab the attention of media today, but rather on the tech breakthroughs that Metzl explains in such clear language. *Hacking Darwin* is essential reading if you want to understand what will be seen as the most important thing happening in this epoch."

—Richard A. Clarke, bestselling author of *Warnings* and
former White House National Security official

"The biosciences revolution is coming at us faster than we realize and what better guide to its possibilities than Jamie Metzl. *Hacking Darwin* is a compelling read ahead of our genetic future, breathtaking on one hand, thought provoking and challenging on the other. A must-read and entertaining guide if you care to know what lies ahead and what to do about it."

—James Manyika, chairman of McKinsey Global Institute

"Jamie Metzl's *Hacking Darwin* should be required reading for business executive, scientists, worriers, and dreamers—for anyone interested in the future of science and humanity. Bit by bit, with wit and grace, he shows how we are rebuilding the very notion of what it means to be human and how we will inevitably be changed in that process."

—Debora Spar, author of *The Baby Business* and Baker Foundation professor at Harvard Business School

"This wildly exciting and highly accessible book kidnaps us on a ride to the future, demonstrating in thrilling detail how our species' ability to transform and improve the genetic makeup of our offspring will explode over the course of our lifetimes and beyond. Even as a scientist working in the field of human longevity and genetics, this book blew my mind. It should be required reading for scientists, doctors, and anybody else whose life and those of their children will be touched by genetic technologies—which is just about all of us. Our world is changing rapidly, and we could have no better guide to the profound opportunities, challenges, and ethical complexities of the genetics revolution than Jamie Metzl."

—Nir Barzilai MD, director of the Institute for Aging Research at the Albert Einstein College of Medicine; director of the Paul F. Glenn Center for the Biology of Human Aging Research and of the National Institutes of Health's (NIH) Nathan Shock Centers of Excellence in the Basic Biology of Aging

"Genetic technologies contain extraordinary promise—and breathtaking challenges—with the potential to change virtually every aspect of our lives. In the pages of *Hacking Darwin*, Jamie Metzl takes us on a critical journey through opportunities and obstacles on humanity's greatest expedition: the exploration and reimagining of what it means to be human."

—Eric Garcetti, Mayor of Los Angeles

"Jamie Metzl has written a personal, funny, unpretentious, and ultimately deeply optimistic book about how the human race will transform itself through genomics. In his enthusiastic embrace of our technological future, Metzl gives voice to the adventure of biological discovery without losing sight of the risks. *Hacking Darwin* is a witty introduction to a wide variety of themes within genomics and is at its most profound when imagining not just the science, but the social and geopolitical reactions that will accrue throughout the world. Metzl writes about real science with the flair of a science fiction writer, and in the process challenges us to begin a complex but necessary conversation about how humanity will evolve."

—Robert C. Green, MD, MPH, professor of medicine (genetics), Brigham and Women's Hospital, Broad Institute, and Harvard Medical School

"Evolution has changed since Darwin's time. We are driving unnatural selection and nonrandom mutation. This gives us the power to alter all species, including ourselves. So the key emerging debate is: Should we redesign humans? Why? How far? How fast? Metzl lays out what we can do now, what we might be able to do, in an engaging, nontechnical way. Let the debate begin."

—Juan Enriquez, coauthor of *Evolving Ourselves: Redesigning the Future of Humanity—One Gene at a Time*

HACKING DARWIN

Also by **JAMIE METZL**

FICTION
Eternal Sonata
Genesis Code
The Depths of the Sea

NONFICTION
Western Responses to Human Rights
Abuses in Cambodia, 1975–1980

HACKING DARWIN

GENETIC ENGINEERING
and the FUTURE *of* HUMANITY

JAMIE METZL

Published by Sourcebooks, Inc.
P.O. Box 4410, Naperville, Illinois 60563-4410
(630) 961-3900
Fax: (630) 961-2168
sourcebooks.com

Library of Congress Cataloging-in-Publication Data

Names: Metzl, Jamie Frederic, author.
Title: Hacking Darwin : genetic engineering and the future of humanity /
   Jamie Metzl.
Description: Naperville, Illinois : Sourcebooks, [2019] | Includes
   bibliographical references.
Identifiers: LCCN 2018041051 | (hardcover : alk. paper)
Subjects: | MESH: Genetic Engineering--ethics | Organisms, Genetically
   Modified | In Vitro Techniques | Human Genetics--trends
Classification: LCC QH437 | NLM QU 550.5.G47 | DDC 576.5--dc23
LC record available at https://lccn.loc.gov/2018041051

Printed and bound in the United States of America.
LSC 10 9 8 7 6 5 4 3 2 1

"Our life is a creation of our mind."

—GAUTAMA BUDDHA

Contents

Introduction

Entering the Genetic Age

"Why are you here?" the young receptionist asked.

It was my first visit to the New York cryobank and I was already feeling a bit uncomfortable.

"I just think it's a good thing for most everyone to do," I said with a shrug. "I lecture around the world on the future of human reproduction and tell anyone who'll listen who wants to have kids they should freeze their eggs or sperm when in their twenties. I'm just a little late."

She raised an eyebrow. *About twenty years late?* "I don't understand. Are you a donor?"

"No."

"Are you going into chemotherapy or having some other medical treatment that could harm your sperm?"

"No."

"Are you in the military about to be deployed?"

"No"

"The only remaining category on my form is *other,*" she concluded after an awkward pause. "Should I put you down for that?"

Already feeling a bit precarious, I didn't want to go into the options I was mulling in my mind. Maybe I'll want to have children someday so may as well store my younger sperm now. Maybe I'll volunteer my sperm to be sent into space when humans starts colonizing the rest of the solar system. Maybe, as I believe, our species is moving toward a genetically altered future in which more of us will conceive our offspring in labs rather than in our beds or the back seats of our cars. Whatever maybes might arise, starting now was the first step.

"Well?" she asked.

I smiled nervously, my mind processing the incredible moment in our evolutionary history where revolutionary new technologies and my own personal biology were intersecting in this antiseptic midtown Manhattan office.

Scientists and theologians can debate whether the first spark of life on our planet sprang from thermal vents on the ocean floor or divine inspiration (or both), but most everyone who believes in science recognizes that around 3.8 billion years ago the first single-cell organisms emerged. These mircroorganisms would have died after one generation if they couldn't find a way to reproduce. But life found a way, and the microbes that started dividing were the ones able to keep their little microbial families going. If each division of these early cells had been an exact copy of the parent, our world would still be occupied solely by these single-cell creatures, and you wouldn't be reading this book. But that's not what happened.

The history of our species is the story of little errors and other changes that kept popping up in the reproduction process.

After a billion years of these small variations created a vast number of slightly different models, one or more of them transformed into simple, multicellular organisms. Still not much by today's standards, these organisms had the potential to introduce even more differences as they reproduced. Some of these variations gave one type of organism or another a small advantage in acquiring food or fending off enemies, providing them the opportunity to live on and mutate more. After two and a half billion years of this, the mutation and competition driving life forward took another miraculous leap with the advent of sexual reproduction.

Sexual reproduction introduced a radical new way of generating diversity when the genetic information of mothers and fathers recombined in novel ways.[1] This incredible process supercharged some of these simple organisms to begin mutating wildly, particularly by around 540 million years ago, into a previously unimaginable diversity of life, including fish. About 200 million years ago, some fish crawled out of the water and evolved into mammals. Around 300,000 years ago, some of those mammals morphed into Homo sapiens, a.k.a. us.

That's basically our evolutionary history. Every one of us is a single-cell organism gone wild through nearly four billion years of random mutation whose ancestors have continually out-competed their competitors in a never-ending cage match for survival. If your ancestors survived and procreated, you are here. If not, you are not. The shorthand name for this is *Darwinian evolution*. It got us to this point. But now the principles of Darwinian evolution are themselves mutating.

From this point onward, much of our mutation will not be random. It will be self-designed.

From this point onward, our selection will not be natural. It will be self-directed.

From this point onward, our species will take active control of our evolutionary process by genetically altering our future offspring into something different from what we are today. We are, in other words, beginning a process of hacking Darwin.

It is an incredible idea with monumental implications.

The current version of our Homo sapiens species was never an evolutionary endpoint but always a stop along the way in our continuous, evolutionary journey. Going forward, we will be driving this process like never before through our technology, hopefully guided by our best values.

If we traveled a thousand years into the past, kidnapped a baby, and brought that baby into our world today, that child would grow up into an adult indistinguishable from everyone else. But if we jumped back into the time machine and went a thousand years from today into the future to do the same, the baby we brought back would be a genetic superhuman by our current standards. He or she would be stronger and smarter than the other children, resistant to many diseases, longer-lived, and have genetic traits today associated with outlier humans like particular forms of genius or with animals like super-keen sensory perceptions. He or she might even carry new traits not yet known in the human or animal worlds but made from the same biological building blocks that have given rise to the great diversity of all life.

"Will the category *other* do?" the receptionist asked, cutting short my reverie.

I took a deep breath. "That's probably the best bet."

"Hmm," she murmured, appearing annoyed I seemed distracted. "And for how long would you like to store?"

"Why don't I start with a hundred years? Let's see how that goes."

She eyed me suspiciously. "I'm sorry, sir, but our storage plans are for one, three, and five years."

My facial expression betrayed my concern. "That's a lot shorter than I was looking for."

"You can always renew."

"That's a lot of renewals," I said with a shrug. "How can I know you'll be around as long as I need?"

"Don't worry. We'll be here. We just renovated our office."

I gulped. Clearly, we were thinking differently about the future of reproduction.

"Please have a seat and fill out these forms," she added, handing me a clipboard, "I'll call you when the doctor is ready."

Sitting nervously in the stiff, red plastic chair under the saccharine Muzak of the white, no-frills waiting room, I filled out the forms and reflected on how I'd arrived to this point. I thought back to the strange series of events that had made me absolutely obsessed with the genetic technologies that will change the evolutionary trajectory of every member of our species, including little old me.

It began when I was working on the White House National Security Council in the second term of the Clinton administration. My then-boss and now close friend Richard Clarke was telling anyone who would listen that terrorism was a major threat to U.S. security and that the United States needed to much more aggressively go after an obscure terrorist named Osama bin Laden. When the 9/11 planes crashed into the Twin Towers, Dick's prophetic and now famous memo on Al-Qaeda was stuck, disregarded, in President Bush's inbox.

Dick always used to say that if everyone in Washington was focusing on one thing, you could be sure there was something far more important being missed. The lesson stuck with me. After leaving the White House, I kept thinking about what were those critically important and under-addressed issues. My mind kept returning to the then nascent revolution in genetics and biotechnology. I became consumed with reading everything I could find and tracking down some of the smartest scientists and thinkers in the world to learn more. When I felt I knew enough to have something to say, I started writing articles on the national security implications of the genetics revolution in foreign policy journals.

One day in early 2008, I got a call out of the blue from a smart and eccentric congressman, Brad Sherman of California. Then chairman of the Subcommittee on Terrorism, Nonproliferation, and Trade of the House Committee on Foreign Affairs, Congressman Sherman told me he'd been thinking a lot about the next generation of terrorist threats. He'd read and appreciated one of my articles and told me he wanted to hold a congressional hearing based on what I'd written. I was honored when he asked me to help frame the event, identify other potential participants, and serve as the lead witness for his prescient June 2008 hearing, "Genetics and Other Human Modification Technologies."

"When our descendants two hundred years from now look back at our present age and ask themselves what were the greatest foreign policy challenges of our time," I asserted in my testimony, "I believe that terrorism, as critically important as it is, will not be on the top of their list. I am here testifying before you today because I believe that how we as Americans and as an international community deal with our new abilities to manage and manipulate our genetic makeup will be."[2]

The attention that came with the congressional testimony gave me confidence that I was on to something important, that I needed to dive deeper into this endlessly fascinating and rapidly changing topic, and that I had a message worth sharing.

I wrote more and more in policy journals and began speaking around the country and world on the future of human genetic engineering. As I continued to learn and engage more, I became increasingly convinced we as a society weren't doing nearly enough to prepare for the coming genetic revolution but worried my message was not getting through. Over time, I began to realize that to more effectively share my message I needed to communicate differently. If my genetics policy lectures weren't breaking through, I needed to reach back into the tool kit I had used once before.

After publishing my first book, an important but largely unread history of the Cambodian genocide filled with thousands of footnotes, I had realized the best vehicle for telling that tale was not a dense historical tome but a story. Telling stories is what we've always done. The tales told in caves and around fires have only now morphed into our novels, movies, and television dramas. My second book, and first novel, *The Depths of the Sea*, explored the tragedy of Cambodian history but this time through a series of intersecting stories of people drawn to the Thai-Cambodian border after the Vietnam War. The first book was a more accurate account of the Cambodian cataclysm, but the novel was far more digestible.

So when facing the challenge years later of trying to bring the critically important issues of the genetics revolution to life beyond my nonfiction writing and speaking, I reverted to my same strategy. In my science fiction novels—*Genesis Code*, which explores the implications of the genetics revolution, and *Eternal Sonata*, a speculation on the future of life extension—I tried to imagine what revolutionary genetic technologies will mean for us on a very human level. I tried to bring people into the story of our genetic future in ways they could more readily absorb.

But then an unexpected thing happened in my book tours. People at my events got a bit excited about the doomsday militias, conniving spymasters, budding romances, and flashbang explosions I'd concocted to give life to my sci-fi world, but their eyes opened widest when I explained the real science of the genetics revolution and what it seemed to mean for us human beings. When I explained the science using the language and storytelling of a novelist, audiences seemed to suddenly understand how the little snippets of scientific information they'd been encountering throughout their daily lives all fit together into the story of our future. I found myself discussing the fiction less and spending more time talking about the very real technology that had the potential to fundamentally transform our species.

The animated conversations I had with people on book tours and at other events challenged me to learn more and inspired me to ask myself even tougher questions about the future of human genetic engineering and my personal relationship to it.

I arrived at my midforties without the children I always assumed I'd eventually have, in part because of my long-standing and not entirely rational faith in science, healthy living, and a positive attitude to check the ravages of time and cruelty of biology. I'm a technology optimist to my core, but as I conjured images of our world to come to my audiences, I found myself wondering if I really believed in the magic of technology as much as I professed.

Did I really believe that the knowledge gained in one hundred and fifty years of genetic science was enough to alter billions of years of our evolutionary biology? Would I really bet that genetic alterations helping make my future child healthier, smarter, and stronger would also make him or her happier? As a student of history, did I not bet that genetically enhanced people might use their advanced capabilities to dominate everyone else like colonial powers have always done? And as the son of a refugee from Nazi Europe, was I really willing to accept the idea that parents could and even should start selecting and engineering their future children based on under-informed genetic theories?

Whatever my answers, one thing was clear: after nearly four billion years of evolution by one set of rules, our species is about to begin evolving by another.

In his farsighted 1865 novel *From the Earth to the Moon*, French novelist Jules Verne described a three-man crew launching themselves in a projectile to the moon and then parachuting home. In 1865, this was a pure work of fantastical science fiction. Very little of the technology that would eventually get humans to the moon a century later had been developed. Imagining a moon landing in 1865 was like imagining humans landing in a different solar system today—it might someday be possible, but we have no real clue how to do it. The science is just not there.

A century later, in 1962, U.S. President John F. Kennedy ascended the podium in Houston to give his now-famous speech announcing that the United States would send a man to the moon by the end of the decade. President Kennedy felt comfortable putting U.S. credibility on the line at the height of the Cold War because in 1962 nearly all the technology that would allow a successful moon landing—the rockets, heat shields, life-support systems, and computers making complex mathematical calculations— already existed. He was neither conjuring a far-off future like Jules Verne nor inventing science fiction. He was drawing very clear inferences from existing technology that only needed some additional tweaks. Nearly every- thing was in place, the realization was inevitable, only the timing was at issue. Seven years later, Neil Armstrong climbed down the Apollo 11 ladder in his "one small step for man, one giant leap for mankind."

For the genetics revolution, now is the equivalent not of 1865 but of 1962. Talk of recasting our species is not speculative science fiction but the logical near-term extension of fast-growing technologies that already exist. We now have all the tools we need to alter the genetic makeup of our species. The science is in place. The realization is inevitable. The only variables are whether this process will fully take off a couple of decades sooner or later and what values we will deploy to guide how the technol- ogy evolves.

Not everyone has heard of Moore's law, the observation that computer- processing power roughly doubles about every two years, but we've all internalized its implications. That's why we expect each new version of our iPhones and laptops to be lighter and do more. But it's becoming increas- ingly clear there is a Moore's law equivalent for understanding and altering all biology, including our own.

We are coming to realize our biology is yet another system of information technology. Our heredity is not magic, we have learned, but code that is increasingly understandable, readable, writable, and hackable. Because of this, we will soon have many of the same expectations for ourselves as we do for our other information technology. We will increasingly see ourselves in many ways *as* IT.

This idea frightens many people and it should. It should also excite us based on its incredible life-affirming possibilities. Regardless of how we feel, the genetic future will arrive far sooner than we are prepared for, building on technologies that already exist.

As a start, we will use the existing technologies of in vitro fertilization (IVF) and informed embryo selection not just to screen out the simplest genetic diseases and select gender, as is currently the case, but also to choose and then alter the genetics of our future children more broadly.

A second, overlapping phase of the human genetic revolution will go a step farther, bumping up the number of eggs available for IVF by inducing large numbers of adult cells like blood or skin cells into stem cells, turning those stem cells into egg cells, and then growing those egg cells into actual eggs.

If and when this process becomes safe for humans, women undergoing IVF will be able to have not just ten or fifteen of their eggs fertilized, but hundreds. Instead of screening the smaller number of their own embryos, these prospective parents would be able to review screens for hundreds or more, supercharging the embryo selection process with big-data analytics.

Many parents will also consider the possibility of not just selecting but of genetically altering their future children. Gene-editing technologies have been around for years, but the recent development of new tools like CRISPR-Cas9 is making it possible to edit the genes of all species, including ours, with far greater precision, speed, flexibility, and affordability than ever before. With CRISPR and tools like it, it will ultimately be scientifically possible to give embryos new traits and capabilities by inserting DNA from other humans, animals, or someday even synthetic sources.

Once parents realize they can use IVF and embryo selection to screen out the risk of many genetic diseases and potentially select for perceived positive traits like higher IQ and even greater extroversion and empathy, more parents will want their children conceived outside the mother. Many will come to see conception through sex as a dangerous and unnecessary risk. Governments and insurance companies will want prospective parents to use IVF and embryo selection to avoid having to pay for lifetimes of care for avoidable and expensive genetic diseases.

With whatever mix of catalysts and first movers, it is almost impossible to believe that our species will forgo chasing advances in technologies that have the potential to eradicate terrible diseases, improve our health, and increase our life spans. We have embraced every new technology—from explosives to nuclear energy to anabolic steroids to plastic surgery—that promises to improve our lives despite their potential downsides, and this will be no exception. The very idea of altering our genetics calls for an enormous dose of humility, but we would be a different species if humility, not hubristic aspiration, had been our guiding principle.

With these tools, we will want to eliminate genetic diseases in the near term, alter and enhance other capabilities in the medium term, and, perhaps, prepare ourselves to live on a hotter Earth, in space, or on other planets in the longer term. Over time, mastering the tools of genetically manipulating ourselves will come to be seen as perhaps the greatest innovation in the history of our species, the key to unlocking an almost unimaginable potential and in many ways an entirely new future.

But that doesn't make all of this any less jarring.

As this revolution unfolds, not everyone will be comfortable with genetic enhancement based on their ideological or religious beliefs or due to real or perceived safety concerns. Life is not just about science and code. It involves mystery and chance and, for some, spirit.

If ours was an ideologically uniform species, this transformation would be challenging. In a world where differences of opinion and belief are so vast and levels of development so disparate, it has the potential, at least if we're not careful, to be cataclysmic.

We'll have to ask, and answer, some truly fundamental questions. Will we use these powerful technologies to expand or limit our humanity? Will the benefits of this science go to the privileged few or will we use these advances to reduce suffering, respect diversity, and promote global health and well-being for everyone? Who has the right to make individual or collective decisions that could ultimately impact the entire human gene pool? And what kind of process, if any, do we need to make the best collective decisions possible about our future evolutionary trajectory as one or possibly more than one species?

There are no easy answers to any of these questions, but every human being needs to be part of the process of grappling with them. We each must see ourselves as President Kennedy stepping to the podium in 1962 Houston, preparing to give our own speech about the future of our species in light of the genetics and biotechnology revolutions. Our collective responses, laundered by our conversations, organizations, civil movements, political structures, and global institutions, will determine in many ways who we are, what we value, and how we move forward. But to be part of that process, we all have an urgent need to educate ourselves on the issues.

"Mr. Metzl, we are ready for you," the receptionist called. I shook my head slightly and looked up, still feeling a little nervous. As the door opened to the back corridor, I stood slowly, paused a moment, then took a deliberate first step forward.

I've written this book to lay out my case for why, even though the human genetic revolution is inevitable and approaching quickly, *how* this revolution plays out is anything *but* inevitable and is, in important ways, up to us. To make the smartest collective decisions about our way forward, we'll need to understand what's happening and what's at stake and bring as many of us as possible into the conversation. This book is my humble effort to jump-start that process.

The door is open for all of us. Whether we like it or not, we are all marching toward it. Our future awaits.

Chapter 1

Where Darwin Meets Mendel

"Raise your hand if you are thinking of having a child more than ten years from now," I asked the large audience of millennials gathered in the sleek Washington, DC, conference hall. About half the audience raised a hand.

I'd been waxing poetic for forty-five minutes about how the coming genetic revolution will transform the way we make babies and ultimately the nature of the babies we make. I'd explained why I believe it is inevitable our species will adopt and embrace our genetically enhanced future, why this was both incredibly exciting and deeply unsettling, and what I thought we needed to do now to try to make sure we can optimize the benefits and minimize the harms of revolutionary genetic technologies.

"If your hand is in the air and you are a woman, you should probably freeze your eggs. If your hand is up and you are a man, I encourage you to freeze your sperm as soon as possible."

The audience eyed me suspiciously.

"No matter how young and fertile you are," I continued, "there's a not-insignificant chance you are going to conceive your children in a laboratory, so you may as well freeze your eggs and sperm now when you are at your biological peak."

A wave of apprehension rolled across the faces of these high-flying young professionals. I could almost feel the conflict brewing. I had struggled for decades with the same question that seemed to be troubling them: How do we balance the magnificent wonder and brutal cruelty of our own biology?

We are all born through a process that feels nothing short of miraculous then immediately begin our never-ending and ultimately losing battle with time, disease, and the elements. We have a strong attraction to what we feel is natural, but our species is defined by our relentless efforts to tame nature. We want our children to be born naturally healthy, but there is practically no limit to how far parents will go in defying nature to save their children from disease.

A young woman in a blue pantsuit raised her hand. "You've just explained where you think the genetic revolution is going and how we should prepare for it, but what about you? Would you genetically engineer your own kids?"

Uncharacteristically, I froze. I'd been writing and lecturing about the future of human reproduction for many years, but somehow the question had never before been asked so directly. I didn't quite know the answer to the woman's question and looked up for a moment to think.

The science of human genetics has advanced so rapidly that all of us are still racing to catch up. When James Watson, Francis Crick, Rosalind Franklin, and Maurice Wilkins identified the double-helix structure of DNA in 1953, they showed how the manual of life is organized like a twisting ladder. Figuring out how to sequence genes just a quarter century later proved that the manual could be read and ever-better understood. Developing tools for precisely editing the genome a few short decades later then allowed scientists to write and rewrite the code of life. Readable, writable, hackable—the scientific advances over the past half-century have turned biology into another form of information technology and humans from indecipherable beings into wetware carriers of our source-code software.

Understanding genetics as IT has led us to increasingly see the genetic variations and mutations causing terrible diseases and enhancing suffering both as the necessary cost of evolutionary diversity and like the annoying bugs interfering with any computer program. Continuing this metaphor, shouldn't we want whatever software updates that might be available to make sure our systems are running optimally?

I felt my thoughts gelling. My eyes regained focus. "If it was safe and I knew I could prevent my child from significant suffering," I said, walking across the stage, "I would do it. If I truly believed I could help my child live a longer, healthier, happier life, I would do it. And if I needed to give my child special capabilities to succeed in a competitive world where most everyone else had advanced capabilities, I would at least think very seriously about it. How about you?"

The woman swayed in her chair. "It's tough," she said, "I hear what you are saying. But something about all this just feels unnatural."

"Let me push you on that," I responded. "What do you mean by *natural?*"

"Probably just things as they are before they've been changed by humans."

"So, is agriculture natural?" I asked. "We've only been doing it for about twelve thousand years."

"It is and it isn't," she said cautiously, starting to recognize nature was a flimsy peg on which to hang an argument.

"Is organic corn natural? Go back nine thousand years and it would be impossible to find anything resembling today's corn. You'd find a wild weed called *teosinte* with a few sad kernels hanging from it. Add millennia of active human manipulations and you get the beautiful, yellow behemoth gracing our picnic tables today. So many of the other fruits and vegetables we eat, even the organic ones from Whole Foods, are in many ways our human creations coming from conscious and selective breeding over millennia. Are they natural?"

"It's a gray area," she conceded, still holding to her original concept of nature.

"Would we be more natural if we lived in hunter-gatherer societies like our ancestors?"

"Probably."

I didn't want to keep pushing but needed to make an essential point. "Would you want to do that?"

An impish smile crossed her face. "Is there room service?"

"So, you are at the Four Seasons and you get a terrible bacteriological infection," I continued. "Would you want to be treated like our ancestors tens of thousands of years ago with incantations and berries or would you want the antibiotics that could save your life?"

"I'll go with the antibiotics," she said.

"Natural?"

"I get your point."

I looked around the room. "We all have deep-seated ideas of what's natural, but much of it isn't that natural at all. It may be what's familiar to us from an earlier time, but we humans have been aggressively altering our world for millennia. And if we have been in the business of altering the biological and other systems around us for so long, must we think of the biology we have inherited from our parents as our destiny? Do we have the right or even the obligation to work out the bugs and software coding errors in the hardware of our and our children's bodies?"

The audience fidgeted.

"If your future child had a terrible disease that you knew would kill him or her, raise your hand if you'd be willing to subject your child to surgery to save his or her life," I pressed on.

All the hands went up.

"If you could prevent your child from having the disease in the first place, would you do that?"

The hands stayed up.

"Keep your hands up if you'd do that by going through IVF and screening your embryos to make sure your future child wasn't at risk."

The hands stayed up.

"How about by safely making one small change to the genes of your child when he or she is just a preimplanted embryo?"

A few hands dropped.

I turned to one of the young men whose hand has dropped, a preppy twentysomething looking like he'd stepped out of the L.L. Bean catalogue. "Can you tell me why?"

"Who are we to start engineering our kids?" he said. "It feels like a slippery slope. Once we start, where do we stop? We could end up with Frankensteins. It makes me nervous."

"That's a very valid point," I said. "It *should* make you nervous. It should make all of us nervous. If you aren't feeling a mix of excitement and fear, you aren't really getting it. Genetic technologies will allow us to do wonderful things that will ease human suffering and unlock potentials we can hardly imagine. New versions of us, Homo sapiens 2.0 and beyond, will use these new capabilities to invent new technologies, explore new worlds, create phenomenal art, and experience an ever-wider range of emotions. But if we don't get things right, the same technologies could divide societies, create oppressive hierarchies between enhanced and unenhanced people, under-mine diversity, lead us to devalue and commodify human life, and even cause major national and international conflict."

"So, who determines where this leads?" another woman asked.

"That will be the most important and consequential question we, individually and collectively, will ask over the coming many years," I said deliberately. "How we answer it will determine who and what we are, where we live and can live, and what is possible for us as people and as a species."

The audience sat up in their seats. I could feel anxiety levels rising in the room.

"We will have to be the ones who figure out where we go with all of this. That's why I'm here speaking with you. Our species as a whole will be making monumental decisions about our genetic future over the coming years. Some of these decisions, like passing laws, will happen on the socie-tal level. But many significant choices will be made by individuals, like each of us figuring out how we want to make babies. Each individual and couple won't feel they are deciding the future of our species, but collec-tively we will be."

That familiar mix of terror, wonder, and confusion I'd come to expect from all of my talks over the years spread across the room.

Then, as always, the hands shot up. Like the seventh graders I'd spoken to in New Jersey, the high rollers in ideas festivals like Google Zeitgeist, Tech

Open Air, and South by Southwest, the experts at Exponential Medicine and the New York Academy of Science, the Stanford and Harvard law students, and the scientists, scholars, and business leaders in conferences around the world, the audience started to understand and internalize the awesome responsibility this historical moment has thrust upon each of us.

It is a responsibility that comes at an incredible inflection point in our history as a species, when our biology and technology are intersecting like never before and are upending some of our most sacred practices and traditions. Like the others, the Washington millennials were starting to grasp that the future of human genetic enhancement wasn't just about making a few changes to our and our children's genes but about creating a new and very different future for our species.

But to understand where we are going, we first need to take a step back to understand where we are coming from.

X

For the first 2.5 billion years of life on earth, our single-cell ancestors reproduced clonally.* One bacterium, for example, would divide into two separate bacteria with the same genetics, and then the process would start over again. This was a great way to do things because you didn't need to waste any time and energy finding a mate. All you had to do was find food and divide and your lineage could go on. The downside was that the clonal reproduction process created a lot of genetic consistency among the single-cell organisms in a given community, limiting the options available for natural selection compared to what would come later.

This consistency, however, was not complete. Bacteria evolved a way to literally grab genes from other bacteria using microscopic harpoons we call *pili*.[1] Still, while clonal reproduction helped bacteria pass on beneficial mutations, it also left some entire colonies at risk when dangers such as

---

*   About 3.5 billion years ago, the first single-cell microbes split into two branches: bacteria and archaea. Some biologists make the case for a third branch, eukarya.

bacteria-infecting viruses arose because the cloned bacteria possessed too many of the same inadequacies in their defense mechanisms. Sexual repro-duction changed that in a big way.

Exact copies in biology are rarely exactly perfect. Although it is impos-sible to pinpoint the exact time, the fossil record suggests that about 1.2 billion years ago one of these simple organisms developed a strange mutation. Rather than just copy themselves or grab a few genes from other microorganisms, they somehow paired with other microbes to create offspring combining the DNA of both parents—et voilà, sex was born, dramatically expanding evolutionary possibilities.

It took more energy to find a partner than it did to clone yourself—there were, by definition, no other potential suitors to contend with. Those on the lookout for optimal partners had to develop new, ever-better capabilities to attract the best mates and fight off competitors. But once a mate was secured, they could more fully and more randomly mix their genetics when procreating—a huge advantage.

Organisms that reproduced sexually had more genetic losers that their clonal forebears, but they also had a far greater possibility of evolving genetic winners. With so many different models of sexually reproducing organ-isms being constantly generated, sexually reproducing species were able to adapt more quickly to changing circumstances, do a better job of fighting off intruders and finding food, and speed up the process of evolutionary change. As one of them, our entire evolutionary history is made of these often-random genetic mutations and variations creating a multitude of new traits, the most useful of which spread across our species. Armed with these differences, our ancestors competed for advantage with each other and the environment around us in a process Darwin called *natural selection*.

Over time, the process of sexual reproduction itself faced evolutionary pressures to which different creatures responded in different ways. Some, like today's salmon, released as many eggs as possible into the world in the hope that some of the eggs will encounter sperm. Releasing thousands of eggs into the holes at the bottom of rivers increased the odds that at least some of them might be fertilized by male sperm, but this approach also eliminated the possibility of parenting. No matter what you may think about your parents, parenting itself confers huge evolutionary advantages.

Rather than sending out huge numbers of eggs, other organisms—our more recent ancestors included—kept the eggs inside the females until fertilization then gestated embryos inside their bodies. If sex were a game of roulette, creatures like salmon puts a chip on every number but creatures like us placed our chips on a few single numbers. By producing fewer offspring than other mammals and keeping them closer to home, our ancestors invested more in raising children, which meant that our kids can develop skills far beyond what a single salmon, hatched and on its own, could ever have.

Sexual reproduction supercharged diversity, creating an ongoing evolutionary arms race. When the salmon won, they reproduced in big numbers but couldn't by definition do anything to raise their kids who were already long gone. We, on the other hand, protected our helpless babies after birth, allowing their brains to keep growing, and nurtured them to provide new skills. Our nature created the evolutionary possibility of our nurture. When we won, we built civilization.

Baked-in sex drives ensured that our ancestors kept sexually reproducing even if they didn't understand, at least on a technical level, much of what was happening. Early civilizations attributed the magic of reproduction to the gods, but our inherently inquisitive brains were hardwired to keep searching for a deeper understanding of the world around us. For millennia, very slow progress was made in understanding our biology, but our knowledge expanded considerably with the advent of the philosophies and tools of the Scientific Revolution.

※

In 1677, the happy Dutchman Antonie van Leeuwenhoek jumped out of bed. Inventor of a far better microscope than anything before, he had already, on his own, peered deeply into bodily fluids of blood, saliva, and tears. This time, however, he recruited his wife. After a sexual encounter, van Leeuwenhoek placed a bit of his ejaculate under his microscope and was amazed to see what he described as "seminal worms" wiggling around "like an eel swimming in water."[2] But what role, he wondered, did these wiggling worms play?

A prominent view in Europe at the time, one that originated with the ancient Greeks, was that male semen contained homunculi, a name for tiny people waiting to start growing. The female body, according to this hypothesis, was like the soil in which a plant seed grows. An alternative belief was that female eggs contained the little mini-mes, whose growth was catalyzed by male semen. A third group of people, presumably mostly dimwits, believed life generated spontaneously like flies emerging from rancid meat.

In the eighteenth century, the brilliant Italian Catholic priest and polymath Lazzaro "Magnifico" Spallanzani crafted an ingenious experiment to test his hypothesis about procreation. Sewing tiny frog pants out of taffeta, he made it impossible for male frogs to pass their "fluids" to the females. Every young person learns this in sex ed today, but in the eighteenth century it was big news that female frogs could not get pregnant when the male sperm was filtered by the pants. When Spallanzani artificially inseminated the female frogs with male frog sperm, the females became pregnant. Sperm, it was now clear, was an essential component of the semen needed to make the females pregnant.[3] *Magnifico!* It took an additional century before scientists figured out that both the male and the female sex cells contribute equally to the fertilized egg.

Learning more about how humans are made then merged with another realization our ancestors had intuitively sensed but never fully understood—the science of heredity.

Drawing of the homunculus by Dutch physicist Nicolas Hartsoeker in 1694.

X

For millennia, our ancestors must have had an inkling about how heredity works. Every time a tall man and a tall woman had a tall child, they received a clue. When a tall man and a tall woman had a short child, they probably got a little confused and, perhaps, the man cast a wary eye at the short, vivacious Casanova living one cave over. Our ancestors used this limited knowledge of heredity to start changing the world around them.

Our nomadic hunter-gatherer ancestors, for example, began noticing that some of the wolves picking through their garbage were friendlier than others. Starting around fifteen thousand years ago, probably in Central Asia, they started breeding those friendlier wolves with each other, eventually creating dogs. Unadulterated by humans, nature on its own devices probably would not have morphed a proud wolf into a yapping Chihuahua, but our ancestors pushed the creation of an entirely new subspecies.

The same human-led domestication process transformed plants. After the vast ice sheets receded almost twelve thousand years ago, our ancestors started replanting particularly useful plants they'd plucked from the wild.[4] Long before Monsanto started genetically engineering seeds, our human ancestors noticed that a few particular plants did something different and more desirable than the others being grown. They figured out that if they planted seeds from these plants, the next generation would more often do the same great thing. Over the coming millennia, this selective breeding process was used to turn wild plants into what we know today as wheat, barley, and peas from the Middle East, rice and millet from China, and squash and corn from Mexico. As humans all around the globe either figured out animal and plant domestication on their own or were exposed to it by others, we increasingly pondered the nature of heredity.

Our ancestors knew how to do heredity but had very little understanding of how it was actually working. For millennia, great human thinkers like Hippocrates and Aristotle in ancient Greece, Charaka in India, and Abu al-Qasim al-Zahrawi and Judah Halevi in Islamic Spain hypothesized about human heredity, but no one had gotten things quite right.

In 1831, an English gentleman explorer with a deep curiosity finagled his way onto a five-year survey voyage along the coasts of Africa, South America, Australia, and New Zealand. A keen observer of detail, Charles Darwin studied his environment carefully, collected huge numbers of specimens, and took meticulous notes. Returning to England in 1836, he spent the next twenty-three years obsessively mulling over his findings and piecing together a powerful hypothesis about how organisms evolve. Darwin recognized his theory would shock Christian morality, so he wanted to be certain he was right before publishing his work. When he learned that a competitor with ideas dangerously close to his own was about to go public, Darwin finally published *On the Origin of Species by Means of Natural Selection* in 1859.

In his masterpiece, Darwin described his theory that all life is related and that species evolve because small changes in heritable traits compete in a process he called *natural selection*. Over time, a species with traits conferring specific advantages in a given environment thrive and repro-duce more than those with less advantageous configurations. If the environment changes, the different traits face different selective pressures in the never-ending process of adaptation and evolution. A highly advan-tageous trait in one environment might become a liability in another, and vice versa. Darwin was spot-on in his theory of evolution but knew little about how heredity actually worked on a molecular level. It took another genius to unlock that mystery.

At the time Darwin's great work was published, an obscure Augustinian friar, Gregor Mendel, was devoting his free time, analytical mind, and careful record keeping to figuring out how traits were passed across generations.

The brilliant son of a peasant farmer, Mendel joined the Augustinian St. Thomas Monastery in Brno (in the modern-day Czech Republic) in 1843. Immediately, he took an active interest in the work already being carried out by other monks trying to better understand how traits were passed on in sheep. Recognizing Mendel's abilities, the abbot sent young Gregor to study physics, chemistry, and zoology at the University of Vienna. After returning from his studies, Mendel convinced the abbot to give him free reign to carry out even more ambitious experiments. Breeding over ten thousand pea plants of twenty-two different varieties between 1856 and 1863, he meticulously recorded how various traits were passed from parent plants to their offspring and painstakingly deduced the laws of heredity that mostly hold true today.

First, Mendel confirmed, each inherited trait is defined by a pair of genes, where one gene is provided by each parent. Second, each trait is determined independent of other traits by the two genes for that trait. Third, if a gene pair has two different genes for the same trait, one form of these genes will always be dominant. Mendel published his revolutionary findings in his seminal 1866 study, "Experiments in Plant Hybridization," and then...nothing happened. Few scientists knew of the paper, which was originally published in the little-read *Proceedings of the Natural History Society of Brünn*. Mendel's incredible work was, for the moment, lost.

But when other scientists exploring the nature of heredity in 1900 stumbled upon tattered copies of Mendel's great paper, the seed of the genetic age was replanted. Ten years later, American biologist Thomas Hunt Morgan proved that the genes Mendel described are organized in structures of molecules called *chromosomes*. Over the coming decades, scientists showed how genetics worked across scores of different organisms. Mendelian genetics became the vehicle underpinning all of life. Combined with Darwinian evolution, it provided the essential keys needed to unlock and then transform all of biology, including our own.

X

All genetic code is made up of very long strands of deoxyribonucleic acid, or DNA, which provides instructions to cells for making proteins. Sexually reproducing species like ours have two paired strands of DNA in the nucleus of nearly all of our cells (our red blood cells don't have nuclei), one from our mother and one from our father. If we were a cake, each of our parents would contribute about half of each ingredient.

But rather than being made up of flour, sugar, and baking soda, our DNA is made up of four different types of molecules called *nucleotides*. These nucleotide "bases" are named *guanine*, *adenine*, *thymine*, and *cytosine*, but each is more commonly referred to by its first letter: *G, A, T,* or *C*. The

G's, A's, T's, and C's are strung together like trains on two sets of parallel tracks, just touching each other. The order of the trains, the sequences of the DNA we call *genes*, creates a unique set of instructions that are delivered by messengers called *ribonucleic acid*, or *RNA*, to the cells for making proteins. Proteins are the real actors in our cells that carry out whatever task has been assigned to them—like becoming a specified type of cell, structuring and regulating our tissues and organs, carrying oxygen, generating biochemical reactions, and growing.

Our human genes are then normally packaged together into twenty-three pairs of DNA strands in our cells—our chromosomes—with each chromosome directing a specific set of functions in our bodies. Humans have about 21,000 genes and 3.2 billion base pairs—points on the genome, the complete set of genes in our bodies—where G's pair with C's and A's pair with T's.

The genes that impact us the most are those providing instructions to our cells to create proteins, but nearly 99 percent of the total DNA does not code for proteins at all. These noncoding genes used to be called *junk DNA* because scientists thought they had no significant biological function. Today we can think of noncoding genes like football players standing on the sideline giving encouragement, tips, and direction to their teammates on the field. These noncoding genes play an important role directing the creation of certain RNA molecules that carry instructions from our genes outside the nucleus and in regulating how the protein-coding genes are expressed.

Each of our cells that has a nucleus contains the blueprint for our whole body, but the result would be chaos if every cell were trying to create the whole person. Instead, our genetic DNA is regulated by a process called *epigenetics* for determining which genes are expressed. A skin cell, for example, contains the blueprint of a liver cell and every other type of cell, but the epigenetic "marks" tell the skin cells to produce skin. In our football team analogy, each player has the entire game plan but only needs to fulfill his or her particular function when instructed to do so.[5]

That's why the single cell of our original fertilized egg can grow into such complicated beings as ourselves. This first cell contains the instructions to generate all the different types of cells, but the cells then begin to specialize in different ways to perform their particular functions. These specialized cells, however, are not independent actors but differentiated parts of an interconnected cellular ecosystem. And just like our organs collaborate with each other within the system of our body, our genes influence each other within the dynamic system of our genome.

This all sounds very complicated, and it is. That's why it's taken hundreds of years to understand how the system works, and we're still only just beginning. But having a recipe and understanding the language of the instructions and the nature of the ingredients is a pretty critical start when baking a cake. Once scientists recognized that genes were the alphabet of the language of life, they still needed to figure out what the letters were saying to be able to read the book. The DNA double helix was the manual made up of letters, but what did the letters say?

Reading the human genome in any significant way was far too difficult for humans alone but not, ultimately, for humans paired with machines. In the mid-1970s, Cambridge scientists Frederick Sanger and Alan Coulson invented an ingenious way to run an electric current through a gel to break up a cell's genome; staining the fragments and sorting the different nucleotides based on length and streaming the gel through a specially designed camera to read the genetic patterns. This first-generation genome sequencing process was slow and expensive but a gargantuan step forward.

By figuring out how to automate this process and better read the flashes of light passing through the DNA "letters," researchers like Lee Hood and Lloyd Smith massively increased both the speed and efficacy of genome sequencing and laid the foundation for another big step forward. When, in 1988, the U.S. National Institutes of Health launched a major initiative to supercharge development of the next generation of these DNA sequencing machines, the stage was set for an even more ambitious initiative to sequence the entire genome.[6]

The Human Genome Project, an audacious, U.S.-led international effort to sequence and map the first human genome, cost 2.7 billion dollars and took thirteen years by the time it was completed in 2003. By then, a private company led by science entrepreneur Craig Venter had pioneered an alternative approach of sequencing the human genome that was less comprehensive but far faster than the government-led effort. Together, these initiatives were truly a giant leap for humankind, and the steps have kept advancing. The launch of companies like San Diego–based Illumina and China's BGI-Shenzhen have turned genome sequencing into a competitive, fast-growing, multibillion-dollar global industry. Next-generation nanopore sequencers that electrically push DNA through tiny holes in proteins to read their contents like ticker tape have the potential to revolutionize gene sequencing even more.[7]

As the technology has gotten more accurate and powerful, costs have gone down dramatically. The chart below gives some indication of how quickly the cost of genome sequencing has decreased over the past decade and a half.

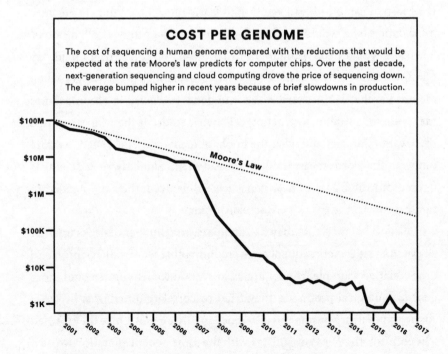

Source: "The Cost of Sequencing a Human Genome," NIH, last modified July 6, 2016, https://www.genome.gov/27565109/the-cost-of-sequencing-a-human-genome/.

Today, sequencing a full genome takes about a day and costs about $700. Illumina CEO Francis deSouza announced in early 2017 that the company expected to be able to sequence a full genome for about $100 in the not-distant future. As the cost of sequencing falls close to the cost of materials required and genome sequencing becomes commoditized, more data will be available at less expense. Because genomics is the ultimate big-data challenge, more and cheaper data will lay a foundation for more and greater discoveries.

But even if sequencing were entirely ubiquitous, commoditized, and free, it wouldn't mean a thing unless scientists were able to understand what the genomes were saying.

X

If a Martian came to Earth wanting to learn about how humans organize information, she would need to figure that we had things called books. Then she would need to figure out that these books had pages made up of words formed by letters. That's the equivalent of what we get when identifying that DNA is organized into genes packaged in chromosomes that code for proteins instructing cells what to do. If the Martian would then want to understand what the books actually said, she'd need to figure out what the words mean and how to read them. Similarly, once scientists figured out the basics of how genes were organized, they still needed to figure out what the genes were actually doing.

The good news is that they had an increasing number of tricks up their sleeve. As researchers sequenced more individual worms, flies, mice, and other relatively simple "model organisms" used to help understand more general biological processes, they tried to correlate differences between similar types of organisms and differences in their genes. Once they formed a hypothesis, they bred organisms with the same genetic mutation to see if the resulting offspring expressed the same trait. Eventually, scientists were able to turn on and off different genes in living animals to observe how specific traits changed as a result. They used advanced computational tools to analyze the interactions of many genes and made broader association studies to analyze ever-larger genetic data sets.

Understanding genetic data sets would be complicated enough if all biology was based on gene expression alone, but it is significantly more complicated. The genome is itself an incredibly complex ecosystem that interacts both with other complex systems inside an organism and with the changing environment around it. A small percentage of traits and diseases result from the expression of single genes, but most come from a group of genes working together and interacting with the broader environment.

No one really knows the exact number, but it has been estimated that hundreds or thousands of genes play a role in determining complex traits like intelligence, height, and personality style. These genes don't act alone. Ribonucleic acid, or RNA, once believed to be merely a messenger between DNA and the protein-making machinery of the cell, is now understood to play an important role in gene expression. The epigenetic marks help determine how genes are expressed. Understanding how these overlapping processes influenced complex genetic traits was way too difficult in the first phase of genomics research, but figuring out the relatively small percentage of traits and diseases caused by single gene mutations was more feasible.

Cystic fibrosis, Huntington's disease, muscular dystrophy, sickle cell disease, and Tay-Sachs are all examples of single-gene mutation diseases, also known as *Mendelian diseases* because they clearly follow Mendel's rules of heredity. Some of these disorders are called *dominant* because a child will need to inherit just one copy of a mutation from a single parent to have the disease. For recessive disorders, like Tay-Sachs, a child would need to inherit the mutation from both parents to be at risk. (In some rare instances people with these mutations don't get the particular disease, most likely because other genes counteract the mutation.) Of the approximately twenty-five thousand Mendelian diseases that have so far been identified, about ten thousand are understood well enough to match a specific gene to a specific disease outcome.[8] Today treatments exist only for around 5 percent of these.

These Mendelian diseases are very rare. Only one out of every thirty thousand people, for example, is born in the United States with cystic fibrosis; one of ten thousand inherits Huntington's disease, and one of 7,250 males inherit Duchenne muscular dystrophy. One of every 365 African American children is born with sickle cell disease, a disease more prominent among groups whose recent ancestors lived in highly malarial areas. Other Mendelian disease can be one in millions or even tens of millions or more.[9]* Many of these diseases cause terrible suffering and even premature death. But because they are so rare, society as a whole generally has less incentive to invest in finding cures to these diseases than in finding cures to more common afflictions, like cancer or heart and lung disease, that impact segments of the population with greater numbers, voice, and political power. Although some new research suggests that variants in Mendelian genes might play a greater role in more common diseases like metastatic prostate cancer, these preliminary findings have so far not shifted the incentive structure.[10]

With so many rare genetic diseases unlikely to receive the attention and resources needed to generate cures, parents and at-risk communities, inspired by the new insights coming from genetic technologies, have started to look for their own ways to protect their future children.

X

Children born with Tay-Sachs, a genetic disease resulting from a single genetic mutation on chromosome 15, often seem fine at birth, but the destruction of their nervous systems begins soon after. By around the age of two, most are experiencing terrible seizures and the decline of mental capacity. Many become blind and nonresponsive. Most die in agony before the age of five. About one in every twenty-seven Ashkenazi Jews are

---

*   A small number of recent studies suggests that mutations putting people at risk for Mendelian conditions are present in about 15 percent of the population. If so, this would change our assessment of the risk these mutations pose and possibly also increase the financial incentives for better understanding and potentially addressing them. Because humans generate tens of additional mutations when generating sperm or egg cells, it is also possible, though less likely, the Mendelian diseases could develop in a child whose parents were not carriers of a particular mutation.

carriers of the Tay-Sachs mutation and hundreds of Jews around the world used to die of the disease each year. Today almost none do, a miracle of both science and social organization.

After scientists in 1969 identified the enzyme associated with Tay-Sachs carriers, a blood test was developed to determine carrier status among prospective parents—and Jewish communities worldwide swung into gear. Jewish community centers and synagogues in the United States, Canada, Europe, Israel, and elsewhere held screenings. Couples in which both prospective parents were carriers were advised to adopt or to get tested after pregnancy. Those mothers carrying embryos with the disease almost always chose to terminate their pregnancies—a painful choice but perhaps less painful than watching their future children die of the disease. The Orthodox Jewish community empowered matchmakers to have marriage candidates genetically tested and steer carriers away from marrying each other.

With the advent of gene sequencing, the genetic mutation responsible for Tay-Sachs was identified in 1985 and multiple mutations of the responsible gene have been identified since. Tay-Sachs is now an exceedingly rare disease among Jewish populations.

In light of the proven benefits of genetic screening for Tay-Sachs, some researchers and policymakers are now calling for what they call "expanded carrier testing" to assess whether other categories of prospective parents have the potential to pass Mendelian diseases and disease-risks to their children.[11]

Genome sequencing and biochemical enzyme level measurement were monumental breakthroughs that began to help prevent the transmission of relatively simple genetic diseases, but genetic analysis alone couldn't transform the way humans make babies unless paired with new options for applying that knowledge.[†] The paired revolutions of in vitro fertilization,

---

†   I use the word *palatable* here because although abortion could address the risk of some Mendelian diseases, it would not be reasonable or desirable to use it to address every Mendelian disease risk.

or IVF, and embryo screening created the mechanism through which genetic analysis could fundamentally transform human baby-making. These revolutions had been a long time coming.

X

Studying rabbit eggs in 1878, more than a century after Spallanzani's experiments with frog condoms, Viennese embryologist Samuel Leopold Schenk—who, coincidentally, studied at the University of Vienna the same time as Gregor Mendel—discovered that when he added sperm to the eggs he isolated in a glass dish, the eggs started dividing. These were the very early years of understanding the reproduction process, but Schenk correctly surmised the eggs were being fertilized. That mammal eggs could be fertilized in a dish suggested that these fertilized eggs could potentially be implanted into the mother and taken to term. In theory, yes; but in practice, not yet. It would take another eighty years until American scientist M. C. Chang successfully impregnated a rabbit with an egg fertilized in a glass dish—or, using the Latin phrase, *in vitro*. But making a bunny was still a far cry from making a human baby. That, too, was coming.

At a historic meeting at the Royal Society of Medicine in London in 1968, biomedical researcher Robert Edwards, one of the world's leading experts on human egg development, approached the leading developer of the surgical process for inspecting a woman's pelvis, obstetrician Patrick Steptoe. Edwards proposed they explore whether human in vitro fertilization could be used to treat infertility. Over the coming decade, the two worked feverishly and published a dazzling string of high-profile scientific papers describing every aspect of what would be required to make human in vitro fertilization possible.

In 1972, Steptoe and Edwards started human trials. Working with nurse Jean Purdy, they carefully extracted eggs from more than one hundred different women, fertilized them with sperm, and then tried to surgically implant the now-fertilized eggs into the prospective mothers. Every one of these efforts failed. In 1976, a woman finally became pregnant with an egg fertilized in vitro, but the pregnancy failed when the early-stage embryo attached outside the uterus. Then, in 1977, Leslie Brown, a homemaker from Bristol, England, entered the clinic. Leslie and her husband, John, a railroad worker, had been trying without success to have a baby for nine years and were desperate.

Leslie's pregnancy took with the first fertilized egg implanted. Nine months later, on July 25, 1978, their healthy baby, Louise, was born. Newspapers across the globe heralded the "baby of the century." When asked just a few months later, a shocking 93 percent of Americans said they had heard about the English baby born from an egg fertilized outside her mother.[12]

Source: "The Birth of the World's First Test-Tube Baby Louise Brown in 1978," *News East West*, July 21, 2013, https://bit.ly/2J3Ymcr.

Although Louise was conceived in a dish, the popular perception was that she, and babies like her, were created in a test tube. The pejorative name *test-tube babies* stuck. Many people, like the majority of Americans polled by Pew that year, had a favorable view of this process.[13] Others felt differently.

Catholic theologians called the test-tube baby process "unnatural" and a "moral abomination" because it did not involve sexual consummation between husband and wife and because the process generated embryos that were not implanted and therefore needed to be discarded.[14] The American Medical Association opposed test-tube baby-making as too aggressive. *Nova* magazine called it "the biggest threat since the atom bomb." Leading conservative bioethicist Leon Kass said it called into question "the idea of the humanness of our human life and the meaning of our embodiment, our sexual being, and our relation to ancestors and descendants."[15] Leslie and John Brown were inundated with hate mail, including blood-spattered parcels containing plastic fetuses.

But, as in many such situations, a process that was once shocking and controversial became more accepted and normalized over time. As the science of "test-tube babies" became less controversial and developed a more technical name, a group of scientists was already imagining the next frontier. Why couldn't cells, they wondered, be taken from an early-stage preimplanted embryo during IVF and then sequenced using ever-advancing sequencing technology?

As early as 1967, IVF pioneer Robert Edwards and his British colleague Richard Gardner described their process for removing a few cells from preimplanted rabbit embryos and screening them under the microscope to determine the sex of the future bunny.[16] In 1990, twelve years after Louise Brown was born, doctors, for the first time, successfully screened a preimplanted human embryo for gender and a few sex-linked and single-gene disorders. This screening process became known as *preimplantation genetic diagnosis,* or PGD. The PGD procedure developed rapidly, particularly for higher-risk prospective mothers. A parallel and related process called *preimplantation genetic screening,* or PGS, was also developed to screen embryos without a known disease risk to assess their chances of thriving. PGD and PGS have more recently been grouped together semantically under the broader umbrella label of *preimplantation genetic testing,* or PGT.

PGT has been around for nearly thirty years, but these are still the early days of this incredibly significant procedure. At first, scientists used PGT primarily to test for the chromosomal abnormalities that can cause miscarriages. It then started to be used to test for a small number of specific single gene disease-causing mutations. Today, PGT is also used to screen for some of the estimated ten thousand single-gene mutation disorders.[17] Unlike prenatal testing of embryos already in the mother's womb, PGT can be carried out on multiple, early-stage fertilized eggs, or *blastocysts,* in a dish.

In most cases, diseases that PGT tests for in these unimplanted embryos are exceedingly rare individually. Collectively, however, they are not. Statistics vary, but recent studies estimate the likelihood of having a traditionally conceived child who carries one of these diseases at about 1 to 2 percent.[18] For the rapidly growing number of screenable single-gene-mutation diseases, the likelihood of a child conceived through IVF and PGT carrying the disease would be massively reduced.[19]

As the number of harmful genetic abnormalities that can be avoided by using IVF and PGT increases, parents will need to weigh the costs and benefits of conceiving children through sex versus in the lab. And while the considerable health and other benefits of conception inside the woman through sex will remain constant—and some small, additional risk associated with the IVF process itself could conceivably be uncovered—the real and perceived health benefits of IVF and embryo screening are likely to increase over time.

Think about all the precautions parents take to protect their children from harm and help them thrive. Mothers swallow prenatal vitamins, douse their own hands and their children's hands in antibacterial sanitizers, have their young kids wear seat belts in cars and helmets on bikes, and serve their children healthy foods. Even though the risk of each different danger varies, modern parents have come to believe that a big part of their job description involves reducing these risks as much as possible and often disdain other parents who make different choices. Most American parents' response to the anti-vaccination movement is a case in point.

When 147 mostly unvaccinated children were infected with measles in 2015, after exposure at Disneyland, the children's parents were roundly condemned for putting hundreds of other kids in danger.[20] While anti-vaccination advocates argue they are doing something "natural" by not vaccinating their children for communicable diseases, it is hard to argue they are actually doing something good.

Vaccinations have saved millions of lives since the first smallpox vaccine was introduced in nineteenth-century England. Repeated studies around the world have clearly proven the safety and overwhelming individual and communal benefits of vaccination.[21] Nevertheless, irrational and uninformed fears of vaccines have persisted. In recent years, celebrities like Jenny McCarthy, Jim Carrey, and Donald Trump[22] have raised scientifically unsupported claims about the dangers of vaccines that have fueled a quadrupling of the number of unvaccinated U.S. children since 2001.[23] This same type of conflict between groups of parents taking action to harness or reject as unnatural scientific advances will also play out with embryo screening.

With the increasing quality and ease of noninvasive prenatal blood tests, many parents already have ever more information about the genetic status of the embryo growing inside the mother. But the anguish of deciding whether to terminate a pregnancy based on genetic abnormalities that could lead to later-stage problems will seem more painful and less beneficial than selecting up front a preimplanted embryo based on statistical probabilities of health.[24]

As the number of single-gene-mutation diseases that can be screened for during IVF and PGT continues to rise, the cost goes down, and the safety of IVF and PGT improves, the value of screening and selecting embryos in the laboratory prior to implantation will increase. At first, parents will balance their faith in reproduction by sex against the benefits of embryo screening. This will not, over time, be a fair fight. With more genetic diseases becoming avoidable, parents who conceive children the old-fashioned way will seem like today's anti-vaccination zealots.

As societal norms about baby-making change, more prospective parents will come to see conception through sex as unnecessarily risky. We'll still have sex for all the wonderful reasons we do now, just not as much for making babies. More parents will want their children to be conceived outside the mother so the embryos can be sequenced, selected, and, in the more distant future, altered.

Although some parents will opt out of this process for ideological reasons or because they get excited together in the back seat of a hovercraft, sexual conception will come with costs. How many people will watch their neighbor's child die from a preventable genetic disease before starting to blame the parents? Will they see these parents as the next "natural" heroes of Disneyland or as ideologues who took unnecessary risks that harmed their children?

A test case for assessing how far prospective parents will go to prevent genetic abnormalities in their offspring has been carried out in Iceland in recent years.

X

Babies with Down syndrome are born with an extra copy of chromosome 21, which can lead to heart defects, developmental and cognitive impairments, increased cancer and mortality risk, and other challenges. Nevertheless, many grow into happy, well-functioning adults who make meaningful contributions to the people around them and society at large. Most people with children, siblings, spouses, or friends with Down syndrome recognize them as the blessing they are.

Since the early 2000s, Icelandic doctors have been required to inform expectant mothers that available screening tests, covered by the national health plan, can indicate with a high degree of certainty whether their future children will have Down syndrome or other genetic disorders. In the decade or so since these tests became available, nearly all the women who received a positive indication for Down syndrome have chosen to terminate their pregnancies. [25] Iceland's termination rate for pregnancies where Down syndrome has been diagnosed largely tracks many other countries. In Australia, China, Denmark, and the United Kingdom, for example, termination rates range from 90 to 98 percent.[26]

With its raging, religion-infused abortion debate, the United States is in many ways an outlier among developed countries. When asked in 2007, only 20 percent of Americans believed parents should be allowed to terminate a pregnancy if their fetus "has a serious, but nonfatal, genetic disease or condition such as Down syndrome."[27] But 67 percent of Americans actually choose to terminate their pregnancies after Down syndrome is diagnosed.[28] The contradictory figures show just how excruciating these decisions can be.*

Critics of universal screening and termination of fetuses indicating Down syndrome raise very valid questions. Who is to say the life of someone with Down syndrome is worth less than anyone else's? What possible moral criteria could be used to make such a determination? These deeply personal questions cut to the core of our very humanity. But these existential questions won't necessarily be what prospective parents will be asking themselves when IVF and preimplantation embryo selection become the norm.

If most mothers and parents in the developed world are already making the often-agonizing decision to terminate existing pregnancies where Down syndrome and other genetic disorders have been diagnosed, imagine what will happen when the choice being made is the less fraught decision about which among the fifteen or so early-stage embryos in a dish to implant? All of these early-stage embryos will be a prospective parent's "natural" children, but only one or two of them can be chosen at a time to take to term.

Doctors performing IVF in fertility clinics are already selecting embryos to reduce the chances of miscarriage. Would we really expect a prospective mother to be agnostic about which among her unimplanted in vitro embryos would be implanted if some might inherit debilitating and life-shortening diseases? Would we want it to be legal for prospective parents to affirmatively choose to implant a baby with Down syndrome or death sentence diseases like Huntington's or Tay-Sachs if it were a choice?

When writing this book, I asked on my Facebook page whether my friends would be willing to edit their preimplanted embryos to give their future children additional traits and capabilities. "As a mom with a kiddo with Down syndrome," an old friend wrote,

---

*   Ohio, Indiana, and North Dakota passed laws in 2017 making it illegal for doctors in those states to perform abortions on embryos because of a Down syndrome diagnosis. If we see an uptick in the number of babies born with Down syndrome in those states over the coming years, we will know that laws like this work. If not, we can assume prospective parents have found alternative ways to have their reproductive wishes expressed.

this is a tough dilemma. If I could have chosen, I believe I would prefer [my son] NOT have Ds [Down syndrome]. But our lives have truly been enriched by this diagnosis. We have a "second family" that is always there to help, no matter what. And he teaches me something new every single day. He's one of the most fun and funny 5-year-olds I know. On the other hand, if I could prevent the struggles he has already faced and those that he will continue to face, I think I would want to prevent that. Every parent wants their kid to be happy and excel at whatever it is that creates that happiness.

Saying that parents would not choose to implant embryos carrying Down syndrome is not at all to suggest that the lives of existing children with abnormalities like this are any less valuable than anyone else's. But because parents around the world are already making the far tougher decision to terminate pregnancies when these kinds of abnormalities appear, it seems likely that parents will affirmatively want to screen out genetic diseases before their pregnancies even begin. Choosing from among preimplanted embryos in a lab will simply seem far less brutal than abortion.

Governments and insurance companies—at least those in jurisdictions with rational health-care systems and where the abortion debate is muted—will also have significant incentives to encourage IVF and preimplantation embryo screening to avoid having to pay for lifetime care for what will come to be seen as avoidable genetic diseases.[†] A relatively simple calculation helps make this point.

Most babies born with early onset genetic diseases spend around three weeks in the neonatal intensive care units of hospitals at an average cost of $3,000 a day or around $60,000 each, and the costs often go up rapidly from there.[29] The mean additional annual cost in the United States for treating cystic fibrosis is $15,571. Because the average life span for people with

---

†   In the irrational U.S. health-care system, where people change insurance providers around every two years, these incentives are decreased.

cystic fibrosis is thirty-seven years, the lifetime additional health-care cost for a person with cystic fibrosis would be nearly $600,000. Considering that there are thirty thousand people in the United States with cystic fibrosis, the total additional annual expenditure on cystic fibrosis alone is about $467 million.[30] The same calculation model arrives at a lifetime total cost of $100 to $150 million for treating the thirty thousand Americans with Huntington's disease and $850 million for treating the roughly two hundred thousand Americans with Down syndrome.[31]* All of these costs are necessary investments in people's loved ones, who deserve every opportunity to reach their potential and enjoy the lives they have, and no price can be set for ameliorating the suffering of even a single human being.

And yet we do make these choices every day through our institutions. If America deployed its full gross domestic product toward curing or treating a particular disease, chances are we could make significant progress. We don't make this investment because we are indifferent to this one disease but because societies require balancing different, valuable interests against each other to function. For preimplantation embryo screening to become accessible to everyone, therefore, the social benefits would need to outweigh the financial and other costs.

If screening all the embryos in the United States for a range of genetic disorders prior to implantation could be done for one dollar less than the total cost of treating for life all the people born with those disorders, society would come out ahead economically while reducing overall levels of pain and suffering. Dividing the total annual cost of treating all these diseases by the total number of babies born each year in the United States gives us a preliminary guess of the point at which every prospective parent could receive IVF and preimplantation embryo screening at no additional cost to society. A rough "back of the envelope" calculation helps make this point.

---

* The more common genetic disorders have higher aggregate costs because more people have them. But costs for treating these disorders can also go down because of economies of scale. Some of the five thousand rarer, treatable single-gene-mutation diseases are extremely expensive because they are so rare that it can take a lot of time and energy to figure out the problem and how best to address it.

Around four million children are born in the United States each year. Assuming that two percent of them are born with genetic diseases, that would mean eighty thousand children.[32] If each of these children had a genetic disease equivalent in cost to the additional roughly $600,000 spent for lifetime care for a person with cystic fibrosis, that would mean spending an additional $48 billion over the next thirty-seven years. If we created an embryo-screening bond to bring forward this future expenditure to apply it today, we'd have about $16,500 to spend on IVF and PGT for each American woman wanting to have a child.† If we included in our calculations the costs of many other genetic and partially genetic diseases that show up later in life—like diabetes, Alzheimer's, and certain cancers—the $16,500 figure would go up farther.

IVF in the United States is an incredibly expensive procedure, usually costing between $12,000 and $30,000 a round. Because couples go through an average of three cycles, these costs quickly become prohibitive to most Americans.[33] But the cost of IVF is far lower in other countries. In Turkey, it costs around $8,500; Britain, $8,000; Spain, $5,600; Mexico, $4,000; Korea, $3,000; and Poland, $1,200.[34] In Israel, where IVF without limit is covered by the national health plan for women under forty-five, national IVF rates are growing at double digits, success rates are high, and costs for medical tourists from abroad are low.[35]

If the cost of IVF and embryo selection in the United States remains high, parents of even moderate means have the ability to go elsewhere for these services. As IVF and embryo screening become the norm, however, prospective parents around the world will increasingly demand these services be covered under their health plans, competition among IVF providers will force prices down, and access and quality and our understanding of what the preimplanted embryos' genes are saying will go up.[36] The consumers, medical providers, and insurance company and/or government payers will all be incentivized to want the same outcome.

Forward-thinking employers are already coming along. In 2014, Apple

---

† This is [48 billion (dollars) divided by 37 (years)] divided by 80,000 parents.

and Facebook announced they would cover the cost of egg freezing for their female employees. This was condemned by a number of prominent women as a ploy to trick women into deferring motherhood and keep working.[37] More people, however, including Facebook's Sheryl Sandberg, saw it as an inevitable step toward empowering women with more reproductive options.[38] Since then, many companies, including Amazon, Google, Intel, Microsoft, Spotify, and Wayfair, have followed suit. More recently, companies such as Starbucks, Facebook, Uber, and NewsCorp have begun covering IVF as part of their employee health plans. According to a survey conducted by FertilityIQ, women working for companies that provide IVF benefits felt a significant bump in their loyalty to their employer.[39] As more employees demand this type of coverage, the best and most competitive employers will provide it.

The intersection of inexpensive and ubiquitous genome sequencing, IVF, embryo selection, and shifting cultural attitudes and financing models will propel more of us to make babies in the future very differently than how we have to date. What we've done in the back seat will itself take the back seat because parents won't get the benefits of genetic selection if they conceive their children through sex. This coming shift away from natural conception would be likely if our understanding of the genome was only advancing linearly, but it is inevitable in light of the exponential progress being made in understanding our genes and how they are expressed.[40]

Assessing how much we can ultimately learn from the extracted and sequenced cells taken from preimplanted embryos during PGT requires us to ask even bigger questions about what genes do and how important they are in determining who we are.

Chapter 2

Climbing the Complexity Ladder

The genetic revolution has provided us with new ways of understanding ourselves that our ancestors could hardly have imagined. Trying to explain to someone twenty thousand years ago that humans are made of code would have been far beyond what their life experiences had prepared them to absorb. But despite our great and often well-founded faith in science, we would be well-advised to maintain the same humble appreciation of the world beyond our grasp as animated our forebears. Even our single-gene-mutation diseases, the clearest and simplest perspective targets for human genetic engineering, should encourage our humility.

Reliably linking single gene mutations to specific genetic diseases represents decades of hard-won progress. But this story is more complicated than it first seems. Because many of the genes linked to particular genetic diseases have been found in people showing symptoms of those diseases, researchers don't know as much as they should about other types of people who might carry similar genetic mutations but who don't get the particular diseases for one reason or another, perhaps because they have some other gene or genes protecting them. That's why it's very likely that the more people—all types of people, not just those showing symptoms of particular diseases—we sequence, the more we will come to recognize the complexity of even seemingly simple genetics. We will learn that we are all genetic mutants in one way or another, carrying mutations that might cause diseases in some but not others.

Our complex and interactive genetics exist within the even greater complexity of our multiple biological systems, the epigenome, transcriptome, proteome, metabolome, microbiome, and virome among them. Our composite individual biology is then embedded within the broader context of our environment.[1]

That's why, after an early stage of euphoria a decade or two ago, many scientists have more recently become more cautious about our time frame for understanding our genetics and the other interacting systems within and around us. We've made enormous advances with inexpensive, fast, and accurate genome sequencing, but our ability to collect data has not yet been matched by an equal ability to understand the data we are collecting. "We've made the mistake of equating the gathering of information," Boston University bioengineer James Collins told *Nature*, "with a corresponding increase in insight and understanding."[2]

As our species always has, however, we balance this justifiable humility about technology with our inherent, Promethean hubris, and for good reason. Each of the complex biological systems within us will be increasingly decodable, our genes foremost among them.

Because tackling the human genome in one swoop is an impossible task, geneticists are working their way up the complexity ladder by trying to understand the biological systems of simpler and faster-breeding model organisms like yeast, fruit flies, roundworms, frogs, mice, and zebra fish, all with many gene and biological systems similar to ours. Because all living beings share a common ancestor, the genetics of these creatures are more or less like that of humans, depending on when we split from them. Humans and fruit flies, for example, share a common ancestor from around seven hundred million years ago. We split with mice, our far closer relatives, a mere eighty million years ago, which explains why both of our species love cheese (just kidding). For this reason, we share 60 percent of our DNA with fruit flies but 92 percent of our DNA with mice.

Unfortunately for them, these relatives of ours have taken the brunt of our genetics research. In the early days, they were bombarded with harmful radiation to mutate their genes and see how various genetic changes led to particular physical outcomes. Today, a wide range of genetic tools are used to knock out genes in model organisms, and laboratories around the world genetically engineer mice and other animals to help study various diseases or traits.* Slowly but consistently, these processes are moving us toward better understanding how complex biological systems like ours function.

For years, researchers like Eric Davidson, a biologist at the California Institute of Technology, have been working to show how the complex biological systems of model organisms can be increasingly understood. Davidson systematically knocked out multiple proteins controlling the expression of genes in sea urchins and monitored how each change altered the other proteins and gene expression. With this information, he and his team are painstakingly developing a dynamic map of how many different proteins and genes interact in an effort to draw basic principles for the overall biological system of the sea urchins. There is still a lot of work to be done, but Davidson describes his work as "a proof of principle that you can understand everything about the system that you want to understand if you get hold of its moving parts."[3]

New genetic tools make it possible to turn on and off multiple genes, but really understanding how genes contribute to complex human traits requires a much more sophisticated process of integration. Genome-wide association studies, or GWAS, are starting to do just that.

Even though all humans are a great deal more genetically alike than not, our relatively small number of genetic differences account for most of our diversity and diseases and so are pretty important. As opposed to the old days of looking for these types of single-gene mutations among groups of

---

*   Around 60 percent of genes connected with human diseases have correlates in flies. Researchers developed hundreds of strains of mice, our closer relatives, to give them all sorts of human diseases and help find cures. Model organism research is absolutely essential to finding cures to human diseases but can cause significant pain for the animals. That's why we need to make sure both that animal experimentation continues and that it is overseen using strong ethical guidelines.

people with the same genetic disease, the GWAS process pores through hundreds, thousands, or even millions of known genetic variations to find differences and patterns that can match with different outcomes.

Once genes are sequenced, the order of the G, A, T, and C bases is translated into a digital file. The GWAS involves a computer algorithm scanning the genomes of large groups of people, looking for genetic variations associated with specific genetic diseases or traits. Each GWAS can look for thousands of these variations (which scientists call *single nucleotide polymorphisms*, or SNPs). The more relevant mutations that are found, the more accurate future studies will become.

To better understand how GWAS and other processes for making sense of immense amounts of genetic data work, imagine trying to comprehend a forest. Imagine that other people have been traveling through the maze of trees and branches for years and have identified thousands of the most significant places in the forest where some of the most important things are happening—perhaps the waterfalls, animal feeding grounds, special plants, etc. We know that these types of sites are significant based on our experience traveling through many other forests. One way to better understand this forest would be to visit each of these high-impact sites and see what's going on. A GWAS does the same within the vast expanse of the genome by seeing what specific genetic markers—ones already flagged as relevant to what we may be looking for—are doing.

Beyond GWAS, newer next generation sequencing (NGS) tools are making it possible for researchers to sequence all of the protein coding genes and then all of the genes in a given genome. Looking at the protein coding genes is like finding a trail that links the most important sites in your forest and allows you to understand how all of the different points along the trail connect to and interact with one another. Sequencing the whole genome is like looking at the entire forest, a bigger and more complicated job but one that ultimately helps us understand the forest far better than just looking at the most important places.

Focusing on such a huge data set as the entire forest or the entire genome is a far more daunting analytical task. It's easier for us to get a sense of a few waterfalls and plants, or of a few target genes, than to understand the broader and more complicated ecosystem of the forest or the genome as a whole. But if we could understand these broader ecosystems, we'd know a great deal more about the forest and, in the case of the human genome, ourselves.

The more we move from looking at how a single genetic mutation causes a disease or trait to how a complex pattern of genes and other systems creates a certain outcome, the less possible it becomes to establish causality using our limited human brains alone. That's why the intersection between the genetics and biotechnology revolutions on the one hand and the artificial intelligence (AI) and big-data analytics revolutions on the other are so critical to our story.

X

The ancient Chinese game of Go, considered by many the world's most complicated board game, has long played a central role in China's culture and strategic thinking. Invented more than 2,500 years ago, the Go board is made up of 361 squares onto which one player places black stones and the other white. Moving in turn, each player tries to encircle the other player's stones to take them off the board. Whoever controls the most territory when the game ends is the winner. To put the complexity of Go into perspective, the average chess move after the first two moves has around 400 options. The average move in Go has around 130,000 options.

Even after IBM's Deep Blue computer defeated chess grand champion Garry Kasparov in 1996, most observers believed it would be many decades before a computer could defeat the world Go champions because Go's mathematical complexity rendered Deep Blue's computational approach useless. But when Google DeepMind's AlphaGo program deployed advanced machine learning capabilities to trounce the Korean and Chinese Go world champions in a series of high-profile competitions in 2016 and 2017, the world took notice.

The AlphaGo program learned how to play Go in part by analyzing hundreds of thousands of digitally recorded human Go matches. In later

2017, DeepMind for Google introduced a new program, AlphaGo Zero, which did not need to study any human games. Instead, the programmers provided the algorithm with the basic rules of Go and instructed it to play against itself to learn the best strategies. Three days later, AlphaGo Zero defeated the original AlphaGo program that had bested the leading grandmasters.

AlphaGo Zero can crush any human Go player because it recognizes layers of patterns in a massive field of data far beyond what any human could hope to achieve on his or her own. The rapid advance of AI has frightened many people, technology entrepreneur Elon Musk and the late Stephen Hawking among them, who are concerned that AI will supplant and someday potentially harm humans.[4] These types of fears are theoretical at our current stage of technological development, but the idea of using AI to begin unlocking the secrets of our biology is not. Today, it is quite clear that AI technology is not supplanting us; it is enhancing us.

Thousands of books have been written about how the information and computing revolutions are transforming the way we store and process information. In the 1880s, punch cards were a major innovation for processing what seemed then like large amounts of data. The magnetic tape was first used in the 1920s to store information and made possible the machines developed to crack Nazi and Japanese secret codes and win the Second World War. Soon after, Hungarian-American genius John Von Neumann laid the foundations of modern computing that underpinned the development of the mainframe, personal computers, and the eventual internet revolution. Now, the connected big data and AI revolutions are allowing us to make increasingly more sense of the growing mountains of data being generated inside and around us.

It is no coincidence that the first word in *big-data analytics* is *big*. More data has been generated in the past two years than in the entirety of human history beforehand, allowing us to do ever-bigger things in a forever accelerating process.[5] This data analytics revolution is massively expanding the problem-solving capacity of our species.

When Thomas Edison was inventing the phonograph, lightbulb,

electrical grid, motion picture camera, and much else in Menlo Park, New Jersey, and he faced a challenge he could not figure out himself, he could speak with a relatively small number of people, perhaps in the hundreds, or read a limited number of books and papers. Today, however, most of our species is networked through the internet; we can glide past problems others have already solved and focus on the new challenges we ourselves are best able to address.

When brilliant people like Edison died, much of their knowledge went with them to the grave. Today, far more of our information and knowledge is captured in our accessible digital records, and the data processing and knowledge-amassing tools we are developing will live in perpetuity. Human death remains an individual and familial tragedy (and just generally sucks), but it has far less impact on the advance of our collective knowledge and our species more generally than it used to. Already today, most of us are in many ways smarter with our smartphones than even great thinkers of the past. We are functionally merging with our breathtaking and fast-improving tools and are, in many significant ways, better off for it.

The big data and machine learning revolutions are helping us figure out all sorts of systems, from urban planning to autonomous vehicles to space travel, but among their most significant impacts will be on our understanding of and ability to manipulate biology. As difficult as Go dominance is, recognizing the patterns of human biology are far more complex. DeepMind's AlphaZero algorithm, a more general version of AlphaGo Zero, isn't just a champion at Go, it also bested leading human champions and inferior algorithms in chess and the complex Japanese strategy game of *shogi*. The rules of these games fed to AlphaZero are simple and straightforward. The "rules" of our biology could well someday be known to us, but today we humans, even working with our AI tools, struggle to figure them out.

To get there, cutting-edge scientists are deploying big data analytical and deep-learning tools to help make more sense of the human genome. Deep-learning software is being used not just to diagnose medical images of breast and other cancers more accurately than human radiologists, but also to synthesize patients' genomic information and electronic medical records to begin diagnosing and even predicting diseases.

Companies around the world are racing to accelerate this process. The innovative Canadian company Deep Genomics, for example, is bringing together AI and genomics to uncover patterns in how diseases work because, in its words, "the future of medicine will rely on artificial intelligence, because biology is too complex for humans to understand."[6] Google and a Chinese company, WuXi NextCODE, recently released highly sophisticated, cloud-based AI systems designed to help make sense of the massive amounts of data coming from genetic sequencing. Boston-based Biogen is looking actively at how quantum computing might supercharge the ability to find meaningful patterns from these massive data sets.[7]

The intersection of AI and genomics will become more powerful as deep-learning techniques become more sophisticated, more and larger data sets of sequenced genomes become available, and our ability to decipher more of the underlying principles of our systems biology grows.

As the genomic data set expands, scientists will use AI tools to better understand how complex genetic patterns can lead to specific outcomes. The real benefit comes from not only sequencing very large numbers of people but also from comparing their genotypes (their genetic makeup) to their phenotypes (how these genes are expressed over the course of their lives). The more sequenced genomes can be matched with detailed life records shared in a common database, the better able we will be to figure out what our genes and other biological systems are doing.

"The world's most valuable resource," *The Economist* wrote in 2017, "is no longer oil, but data."[8] In the case of genomics, it is high-quality data, matching people's biology with the most specific information possible about many other aspects of their lives.[9]

Bringing together these vast sets of genetic data and life records will require relatively uniform electronic health and life records to be analyzed by AI algorithms. Today's diversity of health, medical, and life records systems make the sharing of large pools of genomic data more difficult than it needs to be. In an ideal world, everyone would have their full genome sequenced and all of their personal and medical data recorded accurately in a standardized electronic medical record shareable with researchers in an open network.

In the real world, however, the idea of our most intimate data being made available to people we don't know in a searchable database frightens many of us—for good reason. But different researchers, companies, and governments around the world are exploring different approaches to balancing our collective need for big data pools and our individual desire for data privacy.*

X

Iceland is one of the world's most genetically homogenous societies. Settled by a small number of common ancestors in the ninth century, with relatively few immigrants arriving since, and possessing detailed genealogical, birth, death, and health records going back centuries, the country is an ideal laboratory for genetic research. In 1996, Icelandic neurologist Kári Stefánsson cofounded deCODE Genetics, a company with the ambitious goal of mining the gene pool of Icelanders to better understand and find cures for a range of diseases. To get the data they needed, deCODE convinced Iceland's parliament to grant the company access to national health records and convinced Icelanders, many of whom became company shareholders, to donate their blood to the company.

When Swiss pharmaceutical giant Hoffmann-La Roche bought deCODE for $200 million in 1998, many Icelanders felt betrayed. An ensuing lawsuit denied deCODE access to the national health record system and required each individual to consent to their personal records being shared. But after deCODE and Hoffman-La Roche offered individual Icelanders free access to any drugs developed in the collaboration,

---

* We'll talk more about data privacy later in the book, but I wanted to flag the issue here.

many Icelanders signed back up. Today, deCODE holds 100,000 blood samples and has used its genetic and data pool to discover genes linked to various diseases and even came up with novel treatment for heart attacks.[10] Another pharmaceutical giant, AstraZeneca, announced in early 2018 that it planned to sequence half a million genomes from its own clinical trials by 2026.[11] Governments are also very much involved in efforts to amass large pools of genetic data.

Britain's 100,000 Genomes Project of Genomic England, launched with great fanfare by Prime Minister David Cameron in 2012, is sequencing patients in the country's National Health Service (NHS) with rare diseases and cancer, as well as their families. With the goal of matching genetic information with health records to better understand and advance the treatment of genetic disease, the 100,000 Genomes Project sought to "kick-start the development of a UK genomic industry."[12] Raising the ante, the NHS Genomic Medicine Service announced in October 2018 that all adults with certain cancers and rare diseases would be offered whole genome sequencing with the goal of sequencing five million Britons over the coming five years.

Americans might think the absence of a unified national health system makes this kind of government-led effort more difficult in the United States, but the recently launched U.S. plan is also ambitious. After years of delays, in spring 2018, the U.S. National Institutes of Health began recruiting a targeted million Americans from all socioeconomic, ethnic, and racial groups to submit their sequenced genomes, health records, regular blood samples, and other personal information to the All of Us Research Program.[13] Congress has authorized a $1.45 billion ten-year budget for this program, and enrollment sites are being set up across the United States. If privacy concerns and bureaucratic inertia can be addressed, this initiative could do a lot to push genetic research forward. The U.S. Department of Veterans Affairs also launched its own biobank, the Million Veterans Program, which plans to sequence a million veterans by 2025 to match their genotypes and health and service records.[14]

Creative private-sector models are also emerging that try to balance the societal interest of accessible, big-data pools of genetic information with the interest of many individuals to maintain some level of control over their genetic data. LunaDNA, a young San Diego–based company created by Illumina alumni, is seeking to bring the many small and disparate genetic data sets held by multiple companies and clinics together into a searchable collective by rewarding the individuals willing to share their genetic information with cryptocurrency.[15] This type of approach makes particular sense because people's sequenced genomes, just like their internet search history, will soon have a very significant commercial value whose benefits deserve to be shared with consumers. The Boston-based Personal Genome Project is trying to build an open-source coalition of national genetic data pools.[16]

Perhaps not surprisingly, China has embarked on the most aggressive path for building its big-data genetic pool on an industrial scale. Its recently announced $9 billion, fifteen-year investment to improve national leadership in precision medicine, for example, dwarfs similar initiatives around the world.[17] China's National Development and Reform's thirteenth Five-Year Plan for biotech industry development aims to sequence at least 50 percent of all newborns (including prepregnancy, prenatal, and newborn testing) in China by 2020 and support hundreds of separate projects to sequence genomes and gather clinical data in partnership with local governments and private companies.[18] China is also moving aggressively toward establishing a single, shareable format for all electronic health records across the country and to ensure that privacy protections do not impede access to this data by researchers, companies, and the government.

As a result of all of these types of efforts around the world, it is estimated that up to two billion human genomes might be sequenced within the coming decade.[19] Making sense of this massive amount of data, correlating it to electronic health and life records, and integrating it with large data sets of other human biological systems will require significantly more computing power than we have today, but with the increase in supercomputing capacity around the world there is little doubt we'll eventually get there.[20]

Bringing together so much genetic and personal information in shareable digital databases would be an almost impossible task if each person was making an individual decision about whether to have his or her genome sequenced or not. Instead, most everyone who is born through IVF and embryo selection or visits a doctor's office or hospital at any point in their life will be sequenced as standard procedure—the way people routinely get their pulse tested today—in our collective shift from our system of generalized medicine to the new world of personalized, a.k.a. precision, medicine.

X

Our current medical world is based largely on averages. Not every drug, for example, works for every person, but if it works for even a moderate percentage of people, regulators will often approve it. If you show up in a standard doctor's office with a condition that could be treated with the drug, generally there's a very simple way you find out if it works—by trying it. If you take the common blood thinner Warfarin and it helps, that proves it is right for your biology. If you are among the one in a hundred people for whom Warfarin causes internal bleeding and possibly death, you learn the opposite the hard way.

Generalized medicine was our only way to do things when our understanding of how each individual human being works was low. In the coming world of personalized medicine, this approach will seem the equivalent of leeches. Instead of just seeing a doctor, you will see a doctor paired with an AI agent. Your treatment for ailments from headaches to cancers will be chosen based on how well they work for a person like you. Every person's individual biology—including your gender and age, the status of your microbiome, your metabolic indicators, and your genes—will be the foundation of your medical record and care.

"Doctors have always recognized that every patient is unique," U.S. president Barack Obama said in his 2015 State of the Union Address, "and doctors have always tried to tailor their treatments as best they can to individuals. You can match a blood transfusion to a blood type—that was an important discovery. What if matching a cancer cure to our genetic code was just as easy, just as standard? What if figuring out the right dose of medicine was as simple as taking our temperature?" Soon after, the Obama administration announced its Precision Medicine Initiative designed to "enable a new era of medicine through research, technology, and policies that empower patients, researchers, and providers to work together toward development of individualized care."[21]

Progress toward this goal is being measured one initiative at a time. In 2018, Geisinger Health System (based in Danville, Pennsylvania) announced it was offering free genome sequencing to all patients as part of standard preventative care. Geisinger's preliminary research found actionable findings in 3.5 percent of patients.[22] For the patients, finding potential future dangers can be useful and potentially lifesaving. For the health system, sequencing patients potentially enables better care and could even lead to higher revenue from additional services provided in the short term and could provide savings down the line from preventing more serious conditions. On a societal level, identifying genetic abnormalities early has the potential to make the overall population healthier and reduce the downstream costs of care.*

Although the inefficiencies of health-care systems, the lack of genetics expertise among primary care doctors, and the generally conservative medical cultures around the world could slow the transition, millions then billions of people around the world will eventually have their genomes sequenced as part of the shift toward personalized medicine.[23] Through this process, more and more of our species' genetic, life, and health data will be placed in electronic records, allowing the industrial-scale analysis of our complex biologies.

As the number of people sequenced increases, the cost of sequencing continues to drop, and our computational power to do the necessary big-data analytics climbs, our understanding of complex genetic patterns will grow. Beyond better understanding the impact of an increasing

---

* All of this, of course, is set in the highly irrational U.S. health system, which creates many perverse incentives for everyone.

number of single gene mutations, we will begin to grasp more complex genetic patterns that can lead to polygenic, or multiple gene, conditions like coronary heart disease, diabetes, and hypertension.

This process is already well underway. Researchers at the Broad Institute are building an algorithm and website designed to give people risk scores that assess their genetic predisposition to a range of complex diseases, including coronary artery disease, atrial fibrillation, type 2 diabetes, inflammatory bowel disease, and breast cancer. "[F]or a number of common diseases," according to its letter in *Nature Genetics* in August 2018, "polygenic risk scores can now identify a substantially larger fraction of the population than is found by rare monogenic mutations, at comparable or greater disease risk."[24] Genomic Prediction, a start-up company founded in 2017 by Stephen Hsu, uses these kinds of advanced computer modeling techniques to score percentage risks of a given preimplanted embryo developing a range of complex genetic disorders, including intellectual disabilities, and traits.[25] The company is an early example of the future of expanded PGT. This type of "polygenetic scoring" based on statistical probabilities will be the bridge connecting today's high confidence of predicting many single-gene mutation disorders and our coming ability to predict far more complex disorders influenced by the minimal contribution of many hundreds or thousands of genes.

But there is no possibility that preventing or treating complex genetic disorders will be the endpoint of our genetic journey. It will, in fact, only be the beginning. Pairing assisted reproduction technologies with big-data analytics, machine learning, and AI will increasingly transform not just how we make babies but the nature of the babies we make.

In seeking to unlock the complex genetics of diseases, we will need to understand the bodily systems they impact. To grasp the genetics of cognitive decline, for example, we'll need to understand the genetics of intelligence. To assess the genetics of premature aging, we'll need to understand the broader genetic mechanisms of aging itself. To understand shortness, we'll need to comprehend the biology of height. Seeking to understand genetic abnormalities and climbing the ladder of better understanding of our genetic complexity will, in other words, force us to understand the genetics of being.

And beginning to understand the genetic foundations of our most human characteristics will force us to recognize that the genetics revolution is about much more than our health care. It is about who and what we are today and in the future. Because in a series of incremental steps, it will alter our evolutionary trajectory as a species.

Chapter 3

Decoding Identity

Welcome back to the fertility clinic. The year is 2035.

When you were here ten years ago, you went through the IVF and embryo-screening process to make sure your beautiful daughter was not born with terrible single-gene-mutation diseases. That daughter is now ten and flourishing, giving you a lot of faith in the role embryo screening can play in enhancing a child's health. Watching her and the other children stream out of school at the end of each day, you've also felt sorry for the kids born with genetic abnormalities that could have been avoided had the parents conceived in a lab rather than by selfishly indulging in the random and dangerous conception method of sex.

Raising your daughter over the past ten years, you've paid special attention to the growing stream of discoveries that identify the roles multiple genes play in an increasing number of disorders and traits.

You are back in the same fertility clinic you visited a decade ago, walking from the same waiting room into the same doctor's office, but a lot feels different.

"Doctor," you say, entering her office, "it's nice to see you. I had such a great experience with IVF and embryo selection last time I'd like to do it again." The office, like the waiting room you've just left, feels much more comfortable than a decade ago. The antiseptic white of the walls has been replaced by soft pastels of light blue and lavender. The chairs have gone from industrial to contemporary. The office smells faintly of roses.

The doctor rises to greet you with a warm smile. "That's what everyone says. Can I offer you a cappuccino?"

The question startles you. Since when do they offer barista service at doctors' offices? It seems assisted reproduction has become a competitive customer service business since you were last here. "Can you make it decaf?"

"This is going to be a relatively easy process," the doctor says, then turns her head a moment to dictate your order to her coffee machine. "You were ahead of the game a decade ago when you decided to have us fertilize ten of your eggs."

"I've been sleeping a little easier these past years because I knew the other nine embryos were still frozen. It's hard to imagine I was ever even considering other options. I wasn't sure I even wanted another child, but these past couple of years I haven't been able to escape the feeling that having another is what I need to do."

"And we'll to do everything we can to help turn that feeling into a bubbly, angelic reality. Sugar?"

"Just a teaspoon, please."

"So here is my recommendation," the doctor says, walking over with the teal ceramic mug. "Let's thaw out six of your nine remaining embryos and then extract five cells from each using PGT, just like we did last time."

"Okay," you reply. The doctor hasn't lost the no-nonsense approach you remember from a decade ago.

"Then we'll sequence those cells from each of those six embryos and let you know which of those potential children would be carriers of the single-gene diseases, just like last time."

"Sure," you say confidently, taking a sip of your coffee. It's perfectly brewed. You've been through this process before so feel confident you know what to expect.

"But a lot has also changed since a decade ago," the doctor continues. She leans toward you. "In those early days we could only screen for single-gene-mutation disorders and a few simple traits, like gender and hair and eye color. Now we've learned a lot more about the patterns of multiple genes that can lead to more complex genetic disorders, some of which may not show up until significantly later in life. Because these patterns vary from person to person, and because we still don't fully understand the hugely complex whole genome, we can only see this kind of analysis as predictive. We're no longer dealing with binary outcomes, the on-and-off switch for single gene mutation diseases and disorders we talked about a decade ago. We'll just be able to make percentage predictions, like there's about a 70 percent chance a child born from one embryo, for example, would get disease X before he or she is Y years old. It wouldn't mean that the potential child would get that disease, just that someone with his or her genetics would have that chance of getting it. Of course, this wouldn't account for environmental factors a child would experience after he or she was born. Does that make sense?"

"Yes," you say, a little more cautiously. These are unchartered waters.

"But we are able to make these kinds of predictions for many of the most serious and painful diseases that are influenced by genetics—Alzheimer's, heart disease, some cancers. You won't be able to prevent all those diseases through embryo selection, but you certainly can improve the odds of your future child for delaying or avoiding them."

"A lot really has changed," you say.

"I'm required by law to ask if you want this predictive information. You obviously have the right to refuse it. If you'd like the information, you'll need to sign the form on this tablet."

You think for just a moment, pick up the stylus, and sign. Why wouldn't you want that information? You have to implant one of the embryos anyway. Why not pick the one with the greatest chance to live a healthy life?

"The law also requires me to ask you specifically if you would like to know more about the likelihood that your embryos, if they are implanted and taken to term, would express other non-disease-related traits. It's entirely up to you."

You feel your spine inadvertently stiffen as you begin to comprehend the question. You sense you know the answer but ask anyway. "What types of traits?"

"Some of the most popular are the patterns of genes suggesting a greater chance to live a longer, healthier life."

"That seems like a no-brainer," you say, relieved. Isn't living a long and healthy life what this process is all about, you ask yourself.

"That's good. Some people get nervous the further we get from preventing disease. So many diseases are correlated with age, so if we want to fight disease we also need to defend our children against aging."

"Some of those other people think doctors like you are playing god," you say.

The doctor smiles wistfully at the suggestion. "Some people certainly feel that assisted reproduction is going too far, that we're giving people choices that nature—or whatever deity they trust—didn't want us humans to have. That's why it's so important to find out what each prospective parent is comfortable with. You tell us what works for you, so we can help you achieve it."

"I would've had a lot more qualms when I came in a decade ago, but now I see not selecting the embryos with the greatest chance of a long and healthy life is like taking something away from my future child. It doesn't feel like I'm adding healthy years to his or her life but preventing them from being taken away." You lift the stylus to sign again.

The doctor gives a hand signal for you to wait. "Longevity is just one of the genetic screens. We can also very accurately predict height. Should I go on?"

"I read that tall people have higher incomes and tend to have more self-esteem than short people. Is that true?"

"Most of the studies suggest that."

Do a couple of inches of height mean so much to you that you'd choose a different embryo to get it? But then all of these embryos are your potential natural children, so why not try to pick a taller one if all other things are equal. Picking a taller future child, you explain to yourself, is the same as not picking a shorter one. You take in a deep breath. "Why not? What's the big deal? I'm already selecting for so many other things." The stylus is feeling heavier, but you lift it again.

The doctor again raises her hand gently. "The next screen is IQ," she says softly but with more gravitas. She clearly recognizes the implications of her words.

You've seen the news articles, but something still feels uncomfortable about choosing the potential IQ of your future child. "How accurate is the test?" you ask, stalling for time. "Can we really know something like that?"

"It's all probabilities, but we're getting better at making these kinds of predictions. IQ isn't all about genetics. How you raise and educate your child still means a lot. But IQ is a mostly genetic characteristic, particularly as we age."

"But will my child be happier if he or she has a higher IQ?"

"No one really knows," the doctor says. "IQ is still controversial. Many people say it's culturally biased. But society itself may be culturally biased, so I'm not sure where that leaves us. And there's no denying the correlation between IQ and lots of other important life outcomes."

You inhale again. Do you really want to be in the business of choosing your future child's brain function, you ask yourself. If you don't optimize your child for IQ, will she love you or hate you for it?

"Repeated studies from around the world have shown that people with higher IQs tend to live longer on average than people with lower IQs," she adds.

"How do they know that?" you ask cautiously, a hint of humanist suspicion in your voice.

"Lots of ways. The Scottish government gave IQ tests to all eleven-year-olds in Scotland on a single day in the 1930s. Six decades later, researchers began correlating the IQs from the tests and those children's life experiences. Even when they controlled for social class and lots of other factors, the outcome still showed that the higher IQ kids, on average, lived longer. Scores of additional studies have shown the same thing."

"But IQ isn't just one thing. How can they really know?" you ask, not wanting to reduce the identity of your future child to survey results.

"You're right. IQ is a complicated concept many people reject. Some people even say a high IQ doesn't even make you smart."

"And if someone has a high IQ, does that make them a better artist, a more loyal friend, a more loving father or mother?" you ask.

"Those are all the right questions. The answer to all of them is no. There's no evidence of any of that. But there's a lot of statistical research suggesting that high IQs strongly correlates with success in school, career, wealth creation, and sociability."

You feel yourself relenting in spite of yourself. You know IQ doesn't measure everything and that a human is so much more than a simple IQ score. But are you prepared to reject the concept of IQ out of hand and leave it to your future child to suffer the consequences if you are wrong? That would also be a risk. If you don't select the embryo with the highest IQ it's not at all certain the embryo you select will have a greater genetic predisposition to being a better artist or a more compassionate person. For all you know, those qualities might, like so much else, also be positively correlated with IQ.

But a twinge in your gut also tells you there's something wrong—not about choosing an embryo with a relatively higher IQ than the others but about *not* choosing your higher IQ embryo. This is not the most politically correct thought you've ever had but now, you realize, is a moment for brutal honesty. You squeeze the stylus between your fingers and look up.

"There's more," the doctor continues, a grave look crossing her face. "I need to tell you about some of the latest research on personality styles."

"Personality styles?" you repeat, a lump forming in your throat. What is left of the mystery of being human?

"I imagine you know people in your life who are more extroverted than others."

"Oh yeah," you say, thinking of your sister.

"And people who are more open or more neurotic?"

"I do." Her husband and now, after only six months of puppyhood, their overfed and skittish dog.

"Or even people who are sadistic and cruel."

Your mind shifts to your angry neighbor you saw yesterday kicking his malfunctioning sprinklerbot.

"Personality style has many foundations," the doctor continues, "but genetics is probably the biggest."

"Wait a second," you say, feeling another tug at your humanity. "You're telling me I can select which of these little embryos in your freezer is going to be the next Mother Theresa and which is the next Jeffrey Dahmer?"

The doctor can't seem to decide if you're joking but plays things cautiously. She walks over and sits in the chair beside you. "What I'm saying," she says softly, enunciating each word, "is that we are beginning to understand the genetic patterns underlying different personality styles, and people who want that information when selecting which of their embryos to implant are entitled to it by law—provided they sign a waiver before getting that information."

"A person's personality comes from so many different sources," you say, still trying to hold on to the magical unknown of being human. "How can you reduce all of that to genetics?"

"We can't," the doctor replies, "full stop." She pauses a moment to let her point settle. "But we *can* offer statistical probabilities. If you choose to do so, you have the ability to select the embryo from among your six that has the highest statistical likelihood relative to the others of having whichever personality style you choose."

"Something doesn't seem right about that," you say. "It feels like I'd be ordering my child from Starbucks—light on the milk, extra shot of espresso, three pumps of mocha."

"I'm not here to convince you one way or another," she says, leaning back. "I'm just explaining your options. It's really up to you."

Now your mind is racing. You think back to your own childhood, how surprised your parents were that you were so great at math when neither of them could balance a checkbook. You remember how proud you felt overcoming your shyness to sing in your school talent show. You remember all of the unknown mysteries that unraveled over the course of your life. Would you have felt the same if your parents had selected options for you off a menu? Would they have been as happy when you sang in the talent show or just known you would do it because you had already been genetically optimized for extroversion?

But then again, you counter in your head, all of these embryos are my natural children. One of them will be born into a world where other parents are making these same decisions. If I'm going to invest the coming decades of my life in helping my future child flourish in every way, why wouldn't I pick the embryo with the best shot? You feel your arm quivering. Your hand inadvertently squeezes the stylus ever harder.

To sign or not to sign, that is the question.

X

The more we understand about how our genetics work, the better able we will be to select for more traits—and more genetically complex traits—in our future children.

While we must always be humble about the limits of our knowledge, we are also an aggressive and hubristic species that has always pushed our limits. As our ancestors developed the brain capacity that made group hunting, language, art, and complex social structures possible, we began developing the tools to change the environment around us. When we developed agriculture, cities, and medicine we gave the middle finger to nature as we had found it and it had found us.

But even then, the kinds of choices made in our hypothetical fertility clinic will still require that we figure out how much of what we are stems from our innate biology and how much comes from the broader environment around us. If a disease or a trait is only minimally genetic, selecting it during IVF and PGT makes little difference. For dominant single-gene-mutation diseases like Huntington's disease, or genetic traits like eye color, genes are almost entirely determinative. Lung cancer acquired from a lifetime of smoking, on the other hand, is not.

Fixing our genetic diseases and potentially selecting genetic traits requires that we first figure the extent to which we and each of our disorders or traits are determined by our genes. The process of assessing where biology ends and the environment begins is just another way of describing our age-old debate on the balance between nature and nurture.

Our forebears have debated this for millennia. Plato believed that humans are born with innate knowledge, an assertion later challenged by his star pupil, Aristotle, who argued that knowledge is acquired. Confucius famously wrote in the sixth century BC, "I am not one who was born in the possession of knowledge," casting an unwitting vote for Aristotle. In the seventeenth century, Descartes expressed his Platonic belief that human beings are born with certain innate ideas that undergird our general approach to and attitude toward the world. Hobbes and Locke, on the other hand, believed that lived experience exclusively determined the characteristics of a person. Almost everyone these days would agree that the answer to the nature versus nurture debate is "both," because nature and nurture are both dynamic systems continually interacting. That this is undoubtedly true doesn't make figuring out how genetically determined we are any less important.

This is not just a philosophical question. If we are mostly nature, if our genetics in major ways determine who we are, then fixing a problem or making a change would need to happen on a genetic level. If we are mostly nurture, mostly influenced by the environmental factors around us, then we'd be crazy to think about altering our complex genetics to change outcomes when more benign environmental changes could do the trick.

It's impossible to draw an exact line between nature and nurture, but generations of twin studies have helped scientists understand the role genetics play in influencing who we are. Identical twins separated at birth provide a particularly great opportunity to better understand the role of genetics. The bigger the role genetics play, the greater the likelihood identical twins raised apart will wind up similar to each other.

Identical twins are almost entirely genetic carbon copies of each other at birth; so if humans were 100 percent genetic beings, these twins would remain fully identical throughout their lives. Schizophrenia in twins is a good test case. About half of identical twins share this chronic brain disorder, compared to less than 15 percent of fraternal twins, suggesting that schizophrenia has a significant genetic foundation. But because all identical twins don't share the condition, we know that significant environmental and other nongenetic factors are also involved.

Psychologist and geneticist Thomas Bouchard's multidecade study of twins separated at birth found that identical twins raised separately had about the same chance of sharing personality traits, interests, and attitudes as identical twins raised together.[1] We've all heard the incredible stories of identical twins separated at birth who reunite later in life to find they are shockingly similar.

Identical twins Jim Lewis and Jim Springer were separated when they were only four weeks old. When they met again thirty-nine years later, in 1979, they found they both bit their nails, had constant headaches, smoked the same brand of cigarettes, and drove the same style of car to the same Florida beach. These stories are not mere anecdotes but indicators of a deeper genetic message. Although the twin studies demonstrated that similarities between twins have more to do with genetic and biological factors than with environmental ones, this did not at all negate the critical importance of love, parenting, family, and all types of nurture.[2] Scores of studies like this have been carried out around the world.

In 2015, an enterprising collection of scientists, using findings from most of the twin studies of the previous fifty years, tried to draw conclusions from a mind-boggling collection of 2,748 research publications, exploring 17,804 traits among 14,558,903 pairs of twins in thirty-nine different countries. Using big-data analytics to better pinpoint the balance between genetic and environmental influences, they confirmed that all of the measured human traits are at least partly heritable, but some more than others. At the high end, the measured neurological, heart, personality, ophthalmological, cognitive, and ear-nose-and-throat disorders were found to be mostly genetic. Across all traits, the authors found that the overall average heritability of all of the measured traits was 49 percent.[3] If these findings are correct, humans are about half defined by our genetics. This would be good news for classicists. Plato and Aristotle were both correct.

That we are, overall, probably about half nature and half nurture intuitively feels right. Most parents say they immediately sensed their newborn child had a sunny, anxious, or stormy disposition. Part of this, I am sure, is historical revisionism after a child grows up optimistic, nervous, or belligerent, but part of it is also our recognition that a big part of who we are is based on the biology we inherit. Assuming for the moment that we are about half nature, with some traits being more genetic than others, how far might we go in understanding that genetic and biological part of ourselves?

We know from our intuition and from the research that height is a predominantly genetic trait, but not always. Protein, calcium, and vitamins A and D are essential to helping children grow to their potential. When a disastrous famine struck North Korea in the 1990s, widespread malnutrition stunted the growth of a generation of young North Koreans. The older North Koreans who had come of age prior to the famine were closer to the average South Korean height, but the younger North Koreans who grew up in the 1990s are up to three inches shorter than their South Korean counterparts.[4] This tells us that, whatever your genes might predict, height can be stunted if you don't get the nutrients you need as a child.

If height is mostly genetic, the next step is to figure out which genes have the most to say about how tall we are. There are a small number of single genetic mutations that can make a person very tall or very short. A mutation in the *FBN1* gene, for example, can cause Marfan syndrome, a condition that usually makes people very tall and thin with an extra-long arm span. (Olympic swimmer Michael Phelps and U.S. president Abraham Lincoln both showed symptoms of this mutation.) Achondroplasia, on the other hand, is a mutation in the *FGFR3* gene that causes short-limb dwarfism.

But examples like these of single-gene mutations having a major impact on height are extremely rare. In most cases, height is influenced by hundreds or even thousands of genes as well as by environmental factors like nutrition.[5] Experts believe that about 60 to 80 percent of the difference in height between people is based on genetics.[6] This makes intuitive sense.

A person is generally not tall because he or she has one, single, elongated part, like the neck of a giraffe. Instead, we are tall because each part of us is a little longer. So far, about 800 different genes believed to influence height in one way or another have been identified.

Although the full list of genetic height determinants has not yet been uncovered, Stephen Hsu, a theoretical physicist and vice president for research at Michigan State University, has done incredible work demonstrating how height can be accurately predicted with only the known genetic factors. Drawing on five hundred thousand sequenced genomes from the UK Biobank, Hsu and his collaborators sought to predict the height of people based on their genetics alone. Once these calculations were made, they compared genetic predispositions to the people's actual height. Remarkably, they were able to, on average, predict from their genetic data the actual height of a person within about an inch.[7]

Predicting someone's height from their genetics is useful for catching growth issues early in a child's life, but the stakes of this research are considerably higher. Being able to predict this one complex trait opens the door to the possibility of understanding, predicting, selecting, and ultimately altering any complex trait and many complex and heritable genetic diseases. Hsu and others have already applied similar computational algorithms to predicting heel-bone density, and this approach is being used to predict partial genetic predispositions to diseases like Alzheimer's, ovarian cancer, schizophrenia, and type 1 diabetes, and other polygenic traits, once the sequenced genomes of enough people with each disease and trait have been entered into a shareable database.[8]

It makes sense that predicting a trait that is entirely genetic would be easier if the trait is influenced by just a few genetic markers, rather than hundreds or thousands. That's why we need bigger data pools to predict more complex traits than we do for the genetically simpler ones. If a trait is only part genetic, on the other hand, we can only use genetic data sets to predict its heritable portion. If we have both a sense of how heritable a given trait is and a rough guess of the number of genes influencing the trait, we can then begin to estimate how many people's sequenced genomes and life records would be needed to predict the genetic portion of that trait from genetic data alone. For most common adult chronic diseases, this number is estimated at around a million. For many psychiatric diseases, it is estimated at about one to two million.[9] Most complex and polygenic traits can probably be well predicted from overall data sets in the low millions; the larger the quality data sets, the better.

Height and genetic disease risks are extremely complex and mostly genetic traits, but intelligence, among the most important and complex of all human traits, will be even tougher to tackle.

X

As long as the concept of intelligence has existed, people have debated what it is and how to measure it. General intelligence is extremely difficult to define, even though many people have tried. "The ability to reason, plan, solve problems, think abstractly, comprehend abstract ideas, learn quickly, and learn from experience" captures a lot but also misses a lot.[10]

Intelligence, like all traits, is only valuable within a particular context, and there are about as many different types of intelligence as there are people. No one is smart, beautiful, or strong in absolute terms. Take someone out of their environment and their form of intelligence may be less or more valuable. Albert Einstein was smarter than most of us, but I'm not entirely sure if his type of brilliance would help to find food or water at a time of scarcity.

But although it is politically tempting to argue that any ranking of intelligence constitutes inherent discrimination, our progress as a species demands that we both rank intelligence in general and for specific tasks and that we do our best to ensure that the most capable people are doing the tasks they are best suited for (lest we have our abstract artists running our nuclear power stations). In a world where people's genetic capacities might be matched to their roles, and where general intelligence might be prioritized, we would by definition need to respect many types of intelligence so we can maximally benefit from this diversity (while better realizing our own humanity). Saying that intelligence is diverse, however, cannot prevent our recognition that many of its forms are also hierarchical, especially within particular contexts.

Perhaps more than most every other trait, the battle has long been waged about the heritability of intelligence. In the later 1800s, the English scientist and polymath Sir Francis Galton—whom we will meet again later when we explore the travesty of eugenics—attempted to measure and compare the sensory and other qualities of British noblemen with commoners. This biased effort to demonstrate the gentry's genetic superiority went nowhere but highlights that the very idea of intelligence testing was fraught from the start. A couple of decades later, the French psychologist Alfred Binet designed a series of questions for children to determine which needed special help in school. Based on the belief these questions were ones the average student could answer, those children not able to answer the questions were considered to possess lower than average intelligence.

This idea of having a standard, average intelligence with people either above or below that bar was further developed by the German psychologist William Stern, who fixed that average intelligence quotient, IQ, at 100. If someone had an IQ of 120, they would, according to this model, have a 20 percent higher IQ than the average. An IQ of 80 would be 20 percent lower. American psychologist Charles Spearman recognized that children's cognitive abilities correlated across multiple subjects—that ones who did well in one area tended to do well in others—and created the concept of a *general factor* intelligence.[11]

Over the ensuing years, the concept of IQ spread rapidly around the world and was used in countless organizations, most famously the U.S. military during World War I, to assess a general aptitude applicable to multiple tasks. Although IQ tests probably did help the U.S. military measure certain types of aptitude, the outcome of tests showing IQ disparities between groups, ominously, were also used by some high-profile scholars like Princeton's Carl Brigham to argue that immigration and racial integration would weaken the American gene pool.

Despite this questionable background, IQ, as many studies have shown and as our hypothetical fertility doctor explained, correlates well with a person's health, education level, prosperity, and longevity.[12] Many of the general cognitive abilities most beneficial to our ancestors—including memory, pattern recognition, language ability, and proficiency at math—are positively correlated, and people with high IQs tend to score well on most other cognitive tests. For many critics, however, IQ remains self-validating, inaccurate, disrespectful of difference, racially and socioeconomically biased, dangerous, and, overall, highly questionable.[13]

This debate reached a fever pitch in the United States following the publication of the 1994 book *The Bell Curve: Intelligence and Class Structure in American Life* by Richard Herrnstein and Charles Murray, which described intelligence as the new dividing line in American society. At the start of the book, the authors take the defensible position that general factor cognitive ability can be reliably measured, differs between people, and is somewhere between 40 and 80 percent heritable. Although Herrnstein and Murray recognized that intelligence has both genetic and environmental components, they controversially suggested that genetics explains why some groups score lower than others on IQ tests and that restricting government incentives for poor women to procreate would increase average IQ in the United States.[14] By simultaneously raising the taboo topics of intelligence differences between people and groups, IQ, and race, Herrnstein and Murray smashed into the buzz saw of progressive public opinion.

New York Times columnist Bob Herbert called *The Bell Curve* "a scabrous piece of racial pornography masquerading as serious scholarship."[15] Harvard's Stephen Jay Gould, a longtime critic of the concept of general factor intelligence, argued that other environmental factors like prenatal nutrition, home life, and access to quality education had a more significant influence on a person's intelligence than Herrnstein and Murray accounted for.[16] Other critics attacked the book as reductionist, scientifically sloppy, and dangerously biased.[17]

Even in this highly contentious environment, however, a significant number of researchers sought to defend the principle that general cognitive ability was largely genetic and heritable while trying their best to avoid the clear danger of linking the science of genetics with the contentious politics of race and gender. Fifty-two leading professors published a joint statement in the *Wall Street Journal* in December 1994 asserting that intelligence can be defined and accurately tested, that it is largely genetic and heritable, and that "IQ is strongly related, probably more so than any other single measurable human trait, to many important educational, occupational, economic, and social outcomes."[18] Although Herrnstein and Murray were in many ways flawed messengers making some scurrilous policy recommendations, the idea that IQ had a real, measurable, and heritable genetic component was harder to dispel.

Identical twin and other sibling studies have for decades helped illustrate how much of our IQ is based on our genes versus our experience. The Minnesota Twin Family Study, conducted between 1979 and 1990, followed 137 pairs of twins—81 identical and 56 fraternal—separated early in life and raised apart. Similar to the findings of the researchers on the genetic makeup of height, the Minnesota scientists found that 70 percent of IQ was based on genetics, with only 30 percent resulting from different life experiences, a finding in line with many other twin studies.[19] A more recent literature review of many different studies on the genetics of intelligence estimated IQ to be about 60 percent heritable.[20] Because IQ is a measure of something expressed and realized as we learn and grow (which is why we don't do IQ tests on newborns), the genetic heritability of IQ tends to increase as people get older.[21]

Nevertheless, the very concept of an even partly genetic-based IQ remains sensitive. Questions have repeatedly been raised about whether research like the Minnesota Twin Family Study might have had a socioeconomic bias. In 2003, University of Virginia psychology professor Eric Turkheimer decided to test whether the same percent heritability of intelligence held as true for poorer people as it did for the more middle-class groupings of twins tested in many of the other studies. Interestingly, he found that genetic predictors of IQ were less accurate for children from less advantaged families.[22] One explanation for this could be that IQ tests are fatally biased and flawed. Another might be that negative environmental factors had a greater relative impact than positive ones in these disadvantaged situations—similar to how the food scarcity in North Korea made environmental factors more important for children born in the 1990s than for the rest of the population.

If we accept the findings of the vast majority of studies that IQ is mostly but not entirely genetic, the question then becomes how we can understand the specific genetics underlying IQ. To answer this question, scientists have carried out a large number of studies over recent years trying to identify the thousands of genes influencing IQ. Although only a few genetic variants[23] linked to higher than average intelligence had been identified as of 2016, nearly 200 have been identified since that time.[24] While this progress is astounding, each of the identified genes and gene variants accounts for only a tiny percentage of the estimated IQ differences between people.[25] Identifying a few hundred genes out of the potentially thousands that impact intelligence tells us very little about an actual or potential person's intelligence, in the grand scheme of things.

And yet...

That we are identifying these genes at such a rapid clip at an early stage of the genetic revolution strongly suggests we will find even more as the number of people sequenced, and access to their life records, increases; computing tools and big-data analytic capabilities expand; and more money, people, and educational, business, and governmental resources are being brought to bear.[26] The total number of genes influencing IQ would by definition need to be significantly fewer than twenty-one thousand (the number of genes in the human genome). Our path to identifying the two hundred genes has accelerated exponentially in the past few years, so even a number in the low thousands appears reachable.

Major progress has been made in cracking the genetic code of IQ, and we still have a long way to go, but how much more genetically complicated is IQ than height? Two times more, three time more, five times? It is certainly not ten times more complex.

Steve Hsu's preliminary analysis, and the impressive work of biostatistician Yan Zhang and her colleagues at Johns Hopkins University, suggest that most traits having an average heritability of 50 percent or higher could be predicted with relative accuracy once we have about a million people with their sequenced genomes and life records accessible.[27] Let's say for the moment that human intelligence is five times more complicated than Hsu and Zhang account. That would mean five million sequenced genomes and records. That is a lot of people by today's standards but not compared to the two billion people likely to be sequenced over the coming decade.

Perhaps an even more complicated human trait to understand and quantify is personality style. Personality is among the most intimate human traits, which is why you gulped when the fertility doctor raised it as an option for selection. Although tests like Myers-Briggs have attempted to quantify different personality styles, the process still seems more an art than a science.

A leading theory of personality dating from the 1950s divides human personality styles into five major categories: extroversion, neuroticism, openness, conscientiousness, and agreeableness. Each of us, of course, is somewhere along the spectrum of each of these styles, and measures like these are necessarily relative and subjective. But we can all easily rank our friends and family members from most to least extroverted, neurotic, etc., so the categories are certainly not meaningless.

Per usual, the twin studies also have a lot to say about personality style. "On multiple measures of personality and temperament, occupational and leisure-time interests, and social attitudes," the Minnesota Twin researchers remarkably found, "monozygotic [identical] twins reared apart are about as similar as are monozygotic twins reared together...the effect of being reared in the same home is negligible for many psychological traits."[28]

Twin studies aren't the only way researchers are digging into the genetics of personality style. Scientists at the University of California, San Diego, brought together data to compare hundreds of thousands of people's genomes to information those same people provided in questionnaires about their personality styles in a study reported in 2016 in the journal *Nature Genetics*. When they compared the sequenced genomes of people who described themselves as having similar personality types, the researchers identified six genetic markers that were significantly associated with personality traits. Extraversion was associated with variants in genes *WSD2* and *PCDH15*, and neuroticism with variants in the *L3MBTL2* gene.[29] From just looking at a few points in the genome, the UC San Diego researchers felt they could loosely predict what personality style a person would have.

There is almost no chance that our complex personality styles, in all but a small number of rare cases, are solely or even significantly influenced by single-gene mutations like these, but there is a high likelihood—a certainty even—that more genetic underpinnings of personality will be identified in the coming years.

X

Studies exploring the genetics of height, intelligence, and personality style show that we will increasingly be able to decode the genetic components of even our most intimate, complex, and human traits with growing accuracy. Even if we don't fully uncover the genetics underpinning these traits for many decades, we won't need such a complete understanding to deploy our limited but growing knowledge of genetics in both our health care and reproduction with increasing confidence. We will shift gradually from our generalized health care based on population averages to precision health care based on responding proactively to our genetic predispositions. Perhaps even more significantly, we will begin to integrate these genetic predictions into our reproductive decision-making.

The human implications of having these choices were what made you uncomfortable as the doctor outlined your options in the fertility clinic based on her advanced but incomplete knowledge of how genes function. Even with this limited knowledge, however, each of us will, in one way or another, need to decide whether or not we will sign the doctor's tablet.

Making these kinds of decisions in the 2035 clinic will have a relatively low safety threshold compared to what will come after. Each of these embryos, in 2035, will be the natural and unedited offspring of the genetic parents like you. Even if our understanding of genetics proves completely wrong, all parents using IVF and traditional preimplantation embryo selection will wind up with a child as "natural" as any other.* But at a time when hacking biology is becoming the new operating mode of our species, there is little chance our application of genetic technologies to the reproduction process will end with just selecting from a few preimplanted embryos.

---

*   This calculus would change if it were proven that having children through IVF and PGT was less safe than by sexual reproduction. Because traditional IVF and PGT have been carried out on older and higher pregnancy risk women, making this assessment would require comparing "like with like" potential mothers. So far, there is very little indication that this is the case.

Chapter 4

The End of Sex

I didn't confess everything in the introduction to this book. I stopped the tale prematurely. I hid a word.

The doctor at the cryobank told me that the sprightly receptionist, who also functioned as the nurse, would take me back to the *masturbatorium*. It's a real term.[1]

I blushed a little as she led me through the corridor and handed me a plastic container. She opened the door to a small room that must have once been a broom closet. Sterile white and dominated by a medical supply cabinet, the only indication of the room's purpose was a porn video already playing on the TV. I wondered if someone had profiled me when choosing which DVD to slot in. Did something about me suggest tattoos?

"The magazines are on this rack," she said in a professional tone, pointing to a stack of well-worn magazines. "If you don't like this one, the other DVDs are in this drawer organized by type. Just place your deposit in the container and seal the lid tightly. Leave the specimen on the counter when you are done, and I'll pick it up. Anything else you need?"

I shook my head. A hole to crawl into?

As she stepped out and closed the door behind her, I looked around the strange room feeling ill at ease amid the faint drone of low-volume grunting. Was this really what evolution had in mind for me? There were few environments I could imagine less conducive to the task at hand than this. A dentist's chair came to mind.[2]*

But as awkward as it was, I'd returned to work before my lunch break ended. It really wasn't all that bad. That's because I'm a man.

The average healthy male produces more than 500 billion sperm cells over the course of his life, with between 40 million and 1.2 billion sperm cells being released each ejaculation. Like many other mammalian species, we produce so much sperm because we've been competing with other males for hundreds of millions of years to get our sperm as close as possible to the desired female's eggs. The way our species is built, male sperm is a dime a dozen. That's why the masturbatorium is a thinly disguised broom closet with no technology to speak of other than the television and a Y2K-era DVD player.

But for women the story is different.

Human women are born with about two million egg follicles that will, over time, turn into eggs. Most of these follicles close up before puberty, leaving only about three hundred to four hundred with the potential over the coming few decades to mature into eggs available for fertilization. That's a big difference: about five hundred billion total sperm cells for men, up to four hundred total eggs for women. In contrast to the ease of extracting male sperm, extracting a human female egg is anything but.

When having her eggs retrieved for IVF, a woman must first endure up to five weeks of sex hormone pills and/or injections to make sure she develops a maximal number of follicles and eggs for retrieval. These ovary-stimulating drugs ramp up estrogen levels and not only can cause nausea, bloating, headaches, blurred vision, and hot flashes, but also increase the woman's risk for dangerous blood clots and ovarian hyperstimulation.

Using ultrasound to monitor the development of the woman's eggs attached to the follicle wall in her ovaries, IVF doctors wait until the follicles have reached maturity before anesthetizing the woman for the egg-retrieval procedure. A needle is then passed through the top of the vagina to

---

*   My apologies to any odontophiliacs reading this book!

reach into the ovary to vacuum out the egg, which is then placed in a plastic dish. After the women emerges from sedation and rests, she is usually sent home, where she can expect some cramping for at least a few hours but sometimes longer.

Millions of women endure the pain and potential, though rare, danger of egg retrieval because the promise in their minds of motherhood outweighs this inconvenience. But that doesn't make the process any less difficult.

IVF is not impossible, but it is certainly not easy. It is also less emotionally appealing and more clinical than conception by sex. Conception by sex will always have emotional appeal, and some of us will always trust 3.8 billion years of evolution more than we do our local fertility doctors. But with its mix of costs and benefits, IVF will increasingly compete with sex as the primary way we humans procreate, particularly as it increasingly comes to be seen as safer and more versatile.

Since the birth of Louise Brown in 1978, more than eight million IVF babies have been born with no discernibly different health outcomes than other babies. In the United States, around 1.5 percent of all babies are born with IVF. In Japan, the number has risen to nearly 5 percent.[3] The use of IVF hasn't just helped older or higher risk mothers conceive but has also helped new categories of people, including many gay and lesbian couples, have their own biologically related children.

Starting in the 1980s and '90s, fertility services, donor insemination, surrogacy, and changing social norms paved the way for more gay and lesbian couples to have children of their own in the United States and some other countries. The historic 2015 U.S. Supreme Court decision in *Obergefell v. Hodges* declaring gay marriage protected by the Constitution further facilitated this increasing normalization of the gay family.

With today's technology, gay men who want to have a biologically half-related child generally need to find a donor for the egg that will often be fertilized by one of the men's sperm and then gestated in a surrogate. Lesbian couples need, of course, a male sperm donor. This means that for the estimated 4 percent of the population who self-described to Gallup as being lesbian, gay, bisexual, or transgender (LGBT), assisted reproduction is already not the exception but the norm.[4] As in other areas, the LGBT community stands at the vanguard of social change.

For IVF to become a more common way for all women to conceive, this kind of different thinking about conception will need to go even more mainstream. The IVF model will also need to become easier, less clinical, and less painful for women, a process already well underway.

X

Imagine you are wanting another child in 2045, a decade after your second was born. You are a busy executive with every moment of your day scheduled to the brink. You are proud of your and your partner's ancestry and want to have your own 100 percent biologically related child. Your experience selecting what still feel like positive traits for your second child reaffirmed your data-driven instincts, so this time you don't have many reservations about selecting your embryo prior to implantation with as much genetic information as possible. You are willing to get pregnant because you want to make sure your future child is looked after in utero, even though lots of other people are hiring gestational surrogates and a few are experimenting with synthetic wombs. But you'd still like to make the process as easy as possible, so long as it delivers the maximum benefit. Luckily, you don't have to go through the difficult egg retrieval process because now you can hack your reproductive process more than you'd ever imagined possible.

You are leading your morning staff meeting at work when your assistant comes into your office and places a small Touch Activated Phlebotomy, or TAP, device on your arm. You don't feel a thing and continue with your presentation as the device suctions one hundred milliliters of your blood, the equivalent of about seven tablespoons, into a small receptacle. That's it. In just this moment, you've achieved more than was ever possible with the old IVF egg retrieval model, based on another wonder of human imagination and scientific progress: unlocking the magic of stem cells.

The work of scientists like "Magnifico" Spallanzani showed that human reproduction happens when a male sperm cell fertilizes a female egg cell. If we start as one fertilized egg cell, which then differentiates into the many different types of cells that make up a person, then it logically follows that those early cells must be able to grow out of the initial cell—like plants from a seed. But how?

It was not coincidental that when, in 1908, Russian scientist Alexander A. Maximow described a precursor cell that could develop into any other type of cell, he used a plant analogy. Over the ensuing years, scientists learned more about these amazing "stem cells" that both grow into the many different types of cells it takes to make an organism and then, in a more specialized form, continue regenerating each type of cell to keep the organism going. Stem cells are the seeds that later become the enablers of the branches.

For decades, researchers struggled trying to understand, identify, and isolate stem cells. Then in 1981, British researchers Martin Evans and Matthew Kaufman were the first to extract these stem cells from mouse embryos; a process first carried out on human embryos by two different teams of American researchers seventeen years later. The discovery of human embryonic stem cells launched a tidal wave of anticipation first among scientists and then among the general public.

In 1998, University of Wisconsin developmental biologist Jamie Thomson and Johns Hopkins University's John Gearhart, both leading human embryonic stem cell researchers, described to CNN the many miracles stem cells might unleash. They could, Thomson and Gearhart claimed, be used to repair spinal injuries, heart-muscle cells after heart attacks, and organs damaged by disease or radiation; spur insulin-production to treat diabetes and brain cells to secrete dopamine to treat Parkinson's disease; and to genetically alter blood cells to resist diseases like HIV.[5]

The idea of using embryonic stem cells to deliver these kinds of miracles enthralled many around the world but terrified others. Embryonic stem cells are, as their name implies, derived from embryos. These embryos, in nearly all cases, are unimplanted embryos from fertility clinics. For those believing life begins at conception, these early-stage embryos in petri dishes were people. If so, destroying discarded embryos to support cutting-edge research—even to discover life-saving treatments—was nothing short of murder. America's right-to-life movement quickly targeted stem-cell biology.

"Suddenly, stem cells are everywhere," *Time* magazine reported in July 2001.

> Once relegated to the depths of esoteric health journals, the microscopic clusters have made their way to the nation's front pages...[and] become the political cause du jour in Washington. The debate surrounding the cells threatens to rend traditional alliances, challenging our comprehension of life and leaving some abortion opponents in a very uncomfortable spot: Is it possible to protect the strict boundaries inherent in the "sanctity of life" and still harvest these cells to help the living among us?... [S]ome pro-life advocates have likened using stem cells for research to what Nazi doctors did during World War II. But these cells also hold great promise for millions of ailing patients and their families.[6]

Caught between scientific advancement and his conservative base, and between the conflicting views of different members of his own cabinet, U.S. President George W. Bush split the difference. On August 9, 2001, he announced a ban on federal funding for research involving newly created human embryonic stem-cell lines. This meant that the seventy-one stem-cell lines that met the U.S. government's criteria that were already in use by researchers could still be accessed. Any researchers using new or other human stem-cell lines, however, could not receive U.S. government funding.

The Bush administration's Solomonic effort pleased no one. For advocates of stem-cell research, limiting the number of available stem-cell lines only slowed essential research likely to save and improve countless lives. For many opponents, the use of any embryonic stem-cell lines violated the sanctity of life and was an abomination. Some U.S. researchers decamped to countries like Singapore and the United Kingdom to continue their research with fewer restrictions. Activists in New York and California countered the Bush administration restrictions by creating well-funded, state-based researched support organizations, the New York Stem Cell Foundation and the California Institute for Regenerative Medicine.

But then a discovery in a somewhat obscure Japanese research institute blew open the doors of stem-cell research and transformed our concept of biological plasticity forever.

X

Most of us think of biology as a process that moves forward linearly. We start as a single cell and then grow into complex beings. We are born, we age, then we die. But nature sometimes hides its tricks in plain sight. We wouldn't expect to make perfectly fresh cheese from old milk, but it seems perfectly normal to us that two thirty-year-old adult humans can have a child who is born zero years old, not thirty or sixty. Clearly, our cells already have a way of resetting the clock.

In the 1950s, Oxford biologist John Gurdon was thinking about how certain animal cells seem to rejuvenate themselves. He wondered if cells already specialized—as skin or liver or other types of adult cells—might retain a latent ability to revert to their predifferentiated state. To test this hypothesis, he replaced the nucleus of a frog egg cell with one from a fully mature and specialized frog cell. After a series of these (literally) adulterated egg cells failed to do much of anything after being inseminated with frog sperm, a few, miraculously, became fertilized and healthy tadpoles were created. Gurdon proved that any adult cell, under the right circumstances, had the innate ability to go back in developmental time to become the equivalent of an embryonic stem cell.

Gurdon's incredible discovery opened researchers' minds to the possibility of cells behaving like Benjamin Buttons, but didn't provide a blueprint for how to make that happen. Japanese veterinary researcher Shinya Yamanaka was determined to find it.

As a child in Osaka, Yamanaka enjoyed taking clocks and radios apart and putting them back together. After beginning his career as a surgeon, his frustration that he was unable to heal some of the worst diseases facing his patients inspired him to seek a deeper understanding of cell biology. He went back to school to get his PhD in pharmacology, with a particular focus on mouse genetics. When he learned of Jamie Thomson's work generating human stem cells and John Gurdon's findings on how nuclear DNA can be reprogrammed, he set his heart on deciphering the hidden potencies of cells.

After years of painstaking research, in 2006 Yamanaka and his team discovered that proteins encoded by just four "master genes" could turn back the clock and transform any adult cell back into a stem cell. Just as a stem cell could differentiate itself forward in developmental time into a skin, blood, liver, heart, or any other of the two hundred types of human cells, these Yamanaka factors could revert any skin, blood, liver, or other type of cell back into the equivalent of an embryonic stem cell. He called these *induced pluripotent stem cells*, or iPSCs. Yamanaka deconstructed the cell and showed how the biological clock could tick backward.

Yamanaka and Gurdon won the 2012 Nobel Prize because their discoveries have the potential to revolutionize much of biology—including how we humans create eggs.

In 2012, Japanese cell biologists Katsuhiko Hayashi and Mitinori Saitou announced they had used the Yamanaka factors to reprogram adult mouse skin cells in a dish into iPS cells. They then added more chemicals to turn these stem cells into egg and sperm progenitor cells, the precursors of eggs and sperm. After they placed the same artificial cells into mouse ovaries, the cells matured into eggs. When they put induced sperm precursor into mouse testes, these cells matured into sperm. These induced eggs and sperm were used for mouse IVF, resulting in perfectly healthy baby mice. Even though the success rate for these iPSC-generated sperm and eggs in IVF was extremely low, this was a spectacular breakthrough.[7] Sperm and eggs had been generated from skin cells and used to give birth to healthy mice.

Two years later, in 2014, scientists in England and Israel figured out a way to replicate the same finding, but they made human egg and sperm cells out of adult skin cells in a dish.[8] The efficiency of this process was again extremely low. Then, in September 2018, Saitou and his collaborators announced they had induced egg precursor cells from human blood cells, which were then incubated in tiny ovaries developed from mouse embryonic cells. Even though these induced human egg precursor cells were not mature enough for human fertilization, reaching that point was tantalizingly near. Although the large number of unknowns still makes this procedure not at all safe for humans in the near term, the science is improving fast, and the implications are massive for the future of human reproduction (and your experience in the 2045 clinic).[9]

The average woman who has her eggs extracted yields about fifteen eggs for potential fertilization. Because some of the eggs often prove unfit for fertilization, or because the embryos have one problem or another, a large attrition rate is normal for IVF. The number of early-stage embryos actually available for selection is, therefore, significantly lower than fifteen. But induced stem-cell technology could multiply that number in a very big way.

That one hundred milliliters of blood your assistant took during your meeting contains around three hundred million peripheral blood mononuclear cells, or PBMCs—blood cells with a cell nucleus. Each of these PBMCs (or any other type of adult cell, for that matter) can be transformed by Yamanaka factors into induced stem cells. Then, using the process Hayashi and Saitou developed, each of these millions of stem cells could be turned into egg precursor cells and, ultimately, into eggs.

Now you have hundreds or thousands or even millions of eggs. The father's sperm cells already number in the hundreds of millions. But even if the father's sperm is not at first available or he is infertile, the same process could be used to generate sperm from iPS cells. Put the sperms and eggs together in a dish at the right temperature, or inject the sperm cells into the eggs, and now instead of around fifteen fertilized eggs you have hundreds or thousands. Then the lab can use advanced AI tools to machine-sort the fertilized eggs to identify those with optimal shapes and biologies, the technological version of what embryologists do manually in fertility clinics today. The next step would be to grow these fertilized eggs for five to seven days, until each is about one hundred cells, and then extract five or so cells from each blastocyst and sequence them.

Because the cost of sequencing a genome will soon be largely negligible, sequencing these early-stage embryos will be cheap, easy, and quick. And because low-cost, universal sequencing—paired with electronic medical and life records and big-data analytics—will by then have unlocked far more secrets of the human genome, the information these hundred embryos each provide will be astonishing by today's standards. It will likely be a normal component of the baby-making process then, as you experience in 2045.

In the weeks since the drone messenger picked up your blood sample, you've been living your normal life, but your blood sample has been experiencing a metamorphosis. At the clinic, Yamanaka factors are used to turn your blood cells into induced pluripotent stem cells. These stem cells are induced into egg precursor cells, just like the ones you produced inside your body when you went through IVF a decade ago—only this time many more eggs are being created outside your body than you ever could have produced within.

You've been feeling a growing sense of excitement. You've been talking about the possibilities of a third child for years, but your big day is finally here.

This is a hypothetical, so we can experiment with who we slot into this story in the role of your partner. It could be your husband, but also potentially anybody else. We could imagine a version where two men could have their own 100 percent biologically related babies, using the sperm of one father and eggs induced from an adult cell taken from the other father. Or both the egg and sperm might conceivably be generated from the same man who would become the father and the mother of his child, a scenario seeming more feasible than ever before.

American scientists have already produced viable offspring from two male mice. In November 2018, Chinese scientists announced they had successfully and efficiently harvested stem cells from a mouse egg that they induced into sperm cells used to fertilize the eggs of another female mouse, resulting in healthy mouse pups with half their genetics coming from the first mother and the other half from the second mother. Recent research has pointed to the distinct possibility of generating eggs from 3-D printed ovaries that could be filled with follicles and implanted into a man or transgender woman.

Don't start purchasing your male maternity clothes just yet, but biology is certainly not what it used to be!

On the big day, you wait expectantly in the clinic to be called.

The waiting room has been redecorated since your last visit a decade ago. Settling in to the lush sofa under the soft lights with gentle, spa-style music playing, you feel secure with the choice you've made to stick with this clinic, now part of a larger brand competing in this $50 billion assisted reproduction technologies U.S. business sector.

The door opens, and the doctor comes in to greet you. The office walls are now made up of large screens. She has chosen a beach motif, and as you sit it feels like you are relaxing at a table on the waterfront. The sound of gentle waves soothes you.

You appreciate the effort but have bigger things on your mind. You've now been exposed to scientifically assisted reproduction for two decades and have two wonderful kids—your healthy and robust twenty-year-old, now at college, and your brilliant, artistic, and optimistic ten-year-old at home. This time you don't need any convincing. "Well?" you ask expectantly.

"Okay," she says, "let's get down to it. Your blood sample arrived safely a week ago. We spun your blood in a centrifuge to extract the cells we needed, then reprogrammed them into the stem cells we used to create the egg precursor cells and then your eggs. We decided to fertilize a thousand of your eggs, but then machine sorted those thousand down to the hundred we grew into six-day-old embryos. We then extracted five cells from each of these blastocysts for sequencing and the result is—"

This is the moment you've been waiting for. You hold your breath.

Suddenly, the beach scene goes away and the walls become a massive grid filled with numbers between zero and a hundred. Your eyes scan the room in awe as you stand and look around.

"This may seem a bit confusing," the doctor says, "but the chart shows the probabilities for each of your hundred embryos having the listed outcomes. The Y-axis lists your early-stage embryos from one to a hundred. The X-axis lists the estimated percentage chance each embryo will have a specific condition or trait should that embryo become a person."

There are so many different traits the chart wraps around the entire room. "This is incredible. The science is advancing so fast."

"Welcome to the future of human reproduction," she says.

You walk over and point to the column under the heading *Alzheimer's*. "So, the *90* in this column means that these embryos would have a 90 percent chance of getting Alzheimer's during their lifetimes?"

"Actually no," she replies. "All of the numbers are relative to what's considered the more positive outcome. That 90 percent means that embryo would have a 90 percent chance of *not* getting Alzheimer's if it were selected. The higher numbers are generally what you would likely want for a given trait."

"I see," you say. "And this one would be a great sprinter?"

"He—that embryo's male—would have a genetic proclivity for fast-twitch muscles. And, of course, to be a great sprinter he'd need to have a high score for determination, as well as train, eat well, be exposed to positive role models, et cetera, whatever the genes say."

You scan the walls in awe. And then it hits you. "My mother always used to tell me that I was perfect just as I was."

"I just need to reiterate," the doctor says, "that these are all your natural children, just the same as if you had conceived through sex, just the same as when you went through IVF and embryo selection last time, when the number of eggs was limited by your natural ability to produce them. We've only increased the number of options. All of these qualities, the good ones and the dangerous ones, are all your genetic inheritance. They all reflect the genetics of you and your partner's ancestors going back billions of years. Change is hard for all of us, but the ultimate question people are asking is whether they feel better off with all these choices or better off the old way, where so much was left to chance."

"Am I really just shooting for a high composite score?" you ask, still struggling with the idea that the magic of life can be reduced to a series of percentages on a graph.

"Nature is no fool," the doctor replies thoughtfully. "Evolution isn't random. It just made some trade-offs for us over the years that today don't always seem that great. We have to approach all of this with a great deal of humility."

You look around the room at the cascade of numbers and don't see humility. There's a 10 percent chance some of your embryos won't get Type 1 diabetes, a 20 percent chance some won't get Alzheimer's. Something still doesn't feel right. You've always known that people with genetic disorders were just different. Some of them, like some people with autism, even have superpowers far beyond their so-called normal peers. What does it mean to select these conditions out with a simple nod of the head?

But you also realize these aren't just numbers on a wall. You close your eyes and imagine your future grandchildren holding your third child's hand as her mind deteriorates from Alzheimer's or weeping at the cemetery after her premature death. Would you play Russian roulette with your next child's fate by *not* affirmatively selecting health? Would you even consider not giving your child the best genetic possibilities that your and your partner's genes make possible? If this is hubris, you realize, sign me up. "What's the next step?"

The doctor leans in. "First, we should eliminate all the embryos with high likelihoods of having major diseases. That means the ones with 0 to 50 in the disease-state columns. With so many possible choices, why select embryos so likely to suffer?"

There goes Kafka and Van Gogh, you think to yourself. "And then?"

"There are so many traits influenced in one way or another by genes. The algorithm in this chart weighs each trait by our level of confidence in how important genetic factors are for the trait. For hair and eye color, for example, there's an almost 100 percent certainty that your genes determine the outcome. For some of the other traits, take musical ability for example, we only have a 50 percent confidence the genes we've identified are impactful."

"So where does that leave us?"

"You need to make hard choices in ranking your priorities," the doctor continues. "Picking everything is kind of like picking nothing. If empathy is really important to you, put that in your highest category. If being a good marathon runner with slow-twitch muscles is nice but not that important to you, put that lower so you don't skew the results. Do you follow?"

You hardly answer. Your mind is already transfixed in the range of possible futures playing out on the wall. You take in a deep breath.

"Shall we begin?"

Your mind reaches back to the choices our species has already made.

X

To understand how far selective breeding can take us, we only need to look in our refrigerators.

A wild chicken lays only about one egg a month for some very good reasons. As a species that, unlike salmon, actually raises its young, wild chickens can't produce many more offspring than they are able to rear. Reproduction takes a lot of energy, so laying more eggs would also place more pressure on wild chickens to secure a steady stream of food. And laying a lot of eggs would make wild chickens even more susceptible to lurking predators. For fertile wild females, laying about an egg a month represents a balancing act of evolutionary imperatives played out over millions of years.

Around eight thousand years ago, humans first domesticated chickens from an Asian red jungle fowl called the *Gallus Gallus*. It didn't take Gregor Mendel's meticulous mind or PhDs in genetics for our ancestors to observe which of their newly domesticated chickens were laying the most eggs. It didn't take an IVF clinic or even a galline dating app for farmers to breed the chickens laying the most eggs. Using little science to speak of, our ancestors used selective breeding to amplify a chicken trait that had emerged over millions of years of evolution.

Today, each American consumes an average of 268 eggs per year, amounting to a total U.S. egg consumption of nearly 87 billion a year, produced by around 270 million hens hatching on average one egg a day. Nearly 1.2 trillion eggs are consumed globally each year, produced by a worldwide army of 50 billion domesticated chickens. Many have come to see the large numbers of chicken farms around the world producing these eggs as a dangerous and dirty nuisance.

Imagine what would happen if these domesticated chickens were laying eggs at the same rate as their wild forebears. Instead of 50 billion chickens in farms around world, we'd need around 775 billion chickens. Imagine how people would feel? Earth would become a planet of chickens.

Because the *Gallus Gallus* still runs wild in Southeast Asian forests, it is easy to compare domesticated chickens with their wild ancestors, just as it is easy to compare domesticated dogs and the wolves from which they came. In addition to laying more eggs, domesticated chickens are less aggressive and more social, active, and mobile. As with our other domesticated animals and plants, humans have successfully hacked the chicken.

If we got in our time machine and went back eight thousand years to visit those first chicken-domesticating humans and asked them how many eggs a chicken might lay, they would have been hard-pressed to envision thirty a month. Thirty eggs a month would have sounded then as crazy as a woman today having thirty babies in nine months.

The point here is that biology is significantly more susceptible to hacking than initially meets the eye. We know this intuitively because all of the complex life forms around us evolved from the same origin. But the chicken example suggests that even our own biology might potentially be more manipulable by selective breeding than we are often inclined to believe.

With chickens as a backdrop, now let's look at another organism we will want to hack so it can live longer, healthier, or however we might define it, better.

Homo sapiens like you and me come with time-limited hardware and some very buggy software. Most of the cells in our body turn over, on average, every two years. As we continually copy and recreate ourselves through the magic of our stem cells, small errors start showing up in the genetic code of our new cells. At first this doesn't hurt us much, because we have the machinery inside our cells to fix the errors or keep them from causing damage. But as we get older the number of these errors increases and our ability to fight them declines. We call some of these bugs *aging* and others *cancer*.

As we've come to understand these bugs over the past 150 years, we've also developed some preliminary hacks that allow us to fight back. That's what we're doing every time we treat a genetic disease. When someone gets a bone-marrow transplant, for example, doctors destroy their faulty hematopoietic (blood-forming) cells using radiation and then repopulate the patient with healthy cells from a donor. But if the underlying problem is a heritable genetic disease, the children of even a person cured by a bone-marrow transplant could inherit the same problem.

But what if we wanted to push genetic change like eliminating genetic diseases or enhancing certain traits through our own species as thoroughly as we've pushed increased egg production through chickens? What if we come to see our current biology like our ancestors once saw chickens laying an egg a month—a challenge to be overcome by human ingenuity and selective breeding? The first problem we'd need to overcome would be time, because we humans are just a slow-breeding species.

We enter the world completely helpless and unable to care for ourselves and spend years learning which way is up. At some point in our early teens, some of us begin to recognize there's something different about members of the opposite sex, but we're not quite sure what it is. Terror sets in when some of us realize in our early teens that we need to figure out how the opposite sex works because some day we may need, OMG, to have children with them. We start the process of dating with pimpled awkwardness but then, in our later teens and early twenties, tend to get the hang of it. Having so much fun, we then don't want to end the run prematurely and often aspire to meet enough people to make a smart choice. Finally, at an average age of 27.5 for women and 29.5 for men in the United States and a bit lower worldwide, humans get married. At an average of around twenty-eight in the United States and twenty-six globally, women have their first child, and the process starts again.[10] Twenty-eight years for humans to pass a generation, six months for chickens. That's a big reason why pushing genetic change in chickens can be far quicker than in slow-breeding animals like us.

But what if we could speed up that process to reduce the human generational turnover to about the same six months as a chicken—not by making babies mature faster but by breeding preimplanted embryos with each other? The idea sounds straight out of dystopian science fiction but could someday in the not-too-distant future will be possible.

Imagine we start by taking the mother's blood sample and then induce her peripheral blood mononuclear cells into stem cells, and then into hundreds of eggs. We then fertilize the eggs with the father's sperm and select one embryo (based on whatever criteria). But now, rather than implanting that early-stage embryo in the mother, we instead extract a few cells from it to generate new sex cells. Let's make these new sex cells eggs for the purposes of this hypothetical.

Now imagine that another mother and father went through the exact same process, but instead of inducing eggs from the cells extracted from their preimplanted embryo, they create sperm. If we use this second embryo's sperm to fertilize the first embryo's eggs, then the embryos become the biological parents of their offspring, and the original two sets of mothers and fathers become the grandparents. This grandchild embryo could then be mated with another embryo with an entirely different set of parents and grandparents, now making the original mothers and father great-grandparents (and the original preimplanted embryos grandparents). This process could, theoretically at least, go on forever.

Here's a visual depiction of how this might work:

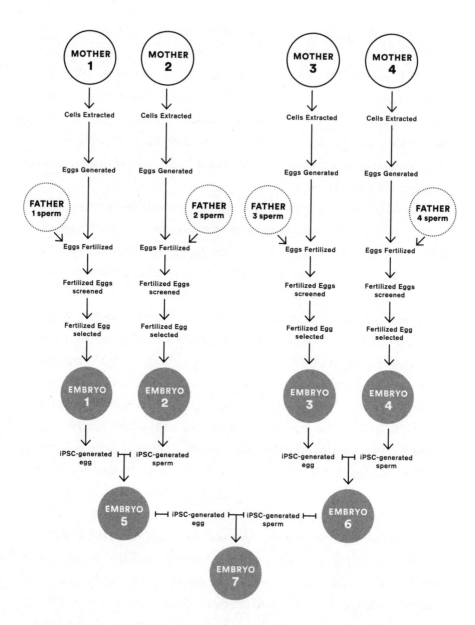

Of course, finding the right embryo to mate with yours and avoiding inbreeding would require a process for preimplanted embryos equivalent to that used by mate-seeking adults today. Perhaps someday an app could be created to help parents find the perfect mate for their precious little preimplanted embryo.

eSwipe: Every embryo
deserves its perfect mate

For technical reasons, this process would take about six months per generation and conceivably could go one forever. Six-month-old embryos could become genetic parents. Year-old frozen embryos (or three-month-old babies) could become grandparents. Using this process would mean we could pass fifty-six human generations over the same twenty-eight years it now takes for one—equivalent of the generational distance between us and Kublai Khan. Rather than three-and-a-half human generations in a century, we could go through two hundred—that number of current human generations would bring us back to just when the wheel was being invented. Sped up by genome sequencing and biological systems analysis driving decisions instead of old-fashioned trial and error, generation change through selective breeding in humans might start seeming as malleable for us as it's been for chickens.

But why might people in the future even consider something as frightening and dehumanizing as breeding human embryos to push genetic changes as we have chickens?

They might not, and this potentially feasible process might remain just a thought experiment. But future generations also might look at the numbers.

X

We've already seen that IQ is influenced by hundreds or even thousands of genes, in most cases each with a relatively small effect. We've also seen that people with higher IQs tend to live longer, earn more, and have more fulfilling relationships than their lower IQ peers, so we know that some parents would conceivably have an incentive to endow their children with the greatest IQ potential possible if they had the choice and believed it was safe.

In their thought-provoking paper, *Embryo Selection for Cognitive Enhancement: Curiosity or Game Changer?*, Oxford professors Carl Shulman and Nick Bostrom attempt to quantify what increase in IQ might be possible based on mating unimplanted embryos with each other.[11]

The IQ of a traditionally conceived child, an *n* of 1, simply has the genetic component of IQ he or she is born with. Because the genetic component of IQ varies among embryos created by the same parents, other than identical twins, we can safely assume the range of IQ options would be greater the larger the number of embryos that might be generated. This means that we would have a greater chance of selecting an unimplanted embryo with a higher IQ if the number of options was larger.

According to Shulman and Bostrom's calculations, the average IQ difference between the highest and lowest IQ of the fifteen or so embryos conceived in IVF, as practiced today, from the same parents would be about twelve points. But if we use the induced stem-cell procedure to start with a hundred fertilized eggs, rather than just ten or fifteen, the average difference between the highest and lowest genetic IQ potentials between these hundred options is estimated at about twenty points. Making a thousand embryos could increase the average differential between the highest and lowest IQ of all the embryos to about twenty-five points.

Twenty-five IQ points may seem like a small return from generating and testing a thousand preimplanted embryos, but that difference would, on average, lead to vastly different life experiences. Einstein was estimated to have an IQ of 160. Arnold Schwarzenegger, not an Einstein but no dummy himself, has an estimated IQ of 135. Enough said.

When we explore the mathematics of mating these already highly selected embryos with each other, the numbers start to look astounding. Shulman and Bostrom estimate that breeding five generations of embryos picked for having the highest IQ among their ten cohorts could create a bump up of sixty-five points in IQ, and a 130-point bump after ten generations. One hundred and thirty IQ points is the difference between an Einstein and a person with severe intellectual disabilities who requires constant care. In the other direction, it's the difference between an Einstein and someone with, by far, the highest IQ ever recorded.

Stephen Hsu is even more optimistic about how much of a bump might be possible. Normally, some human genes have either a positive or a negative effect on intelligence, resulting in a normal curve of intelligence. If each of these genes is tweaked to have a positive effect, then the aggregate effect could be a human with a much higher level of intelligence than the 100 standard deviation IQ average. Hsu believes that because we will likely be able to identify most of the genes associated with intelligence within a decade, parents in the future might be able to select between preimplanted embryos based on the expression of genes associated with intelligence to create a super-intelligent person with an IQ of 1,000.

Hidden in these calculations of multiples and averages is the increasing possibility that future children with truly outstanding outlier capabilities—such as a genius for math, making music, or writing advanced computer code—might revolutionize our ways of thinking even more than our greatest geniuses like Einstein, Confucius, Marie Curie, Isaac Newton, Shakespeare, and David Hasselhoff. "We can imagine savant-like capabilities," Hsu writes, "that in a maximal type, might be present all at once: nearly perfect recall of images and language; super-fast thinking and calculation; powerful geometric visualization, even in higher dimensions; the ability to execute multiple analyses or trains of thought in parallel at the same time; the list goes on."[12]

Of course, we have no idea what an IQ of 1,000 would mean for a person. Evolution is full of redundancies, protections, and trade-offs. It is very possible, perhaps even likely, that a person engineered to have a 1,000 IQ might be driven crazy, become a dangerous sociopath, or develop some type of neurological malady we've never seen. It would be extremely difficult to know what harmful mutations were being pushed across multiple generations of embryos mated before an actual human child was born. Creating super-intelligent humans would, of course, also have massive social and ethical implications.

It certainly sounds crazy and frightens many people to imagine that future humans could have IQs so far beyond history's greatest geniuses, but is it not crazy to believe our current IQ range makes us kind of like the wild chickens popping out a humble egg a month. It may be that our evolution to date has been so optimized for intelligence in balance with our other survival imperatives that such big IQ gains would require physically larger brains not possible within the space constraints of the craniums we now have.*

If so, one solution to this hypothetical problem would be to optimize future children for larger craniums. Futurists like Ray Kurzweil predict this limitation on what our biological brains can do will force us to merge with our machines, so our mental capacities can expand exponentially, perhaps in a computing cloud, without bumping into any physical space constraints. Even then, our species might need some super-biological geniuses with IQs far beyond today's normal to write the code that guides our AI (at least until the AI starts writing its own code) or helps make this expansion of understanding and consciousness available to everyone.†

Human women will not be selecting from among hundreds of preimplanted embryos for at least a couple of decades, and human embryos will not be mating for significantly longer. But if and when this becomes possible, we'll face a host of incredibly thorny ethical questions. Would the responsibility of a mother giving birth to her great-great-great-great-grandson be any different than to a direct biological child? Are unimplanted embryos in a dish entitled to any rights? Is mating embryos without their consent a form of slavery?

But as strange as these types of hacking human reproduction seem today, any child born from this process would have 100 percent unadulterated human DNA. If that doesn't seem like much of a consolation, it's because we haven't yet begun to discuss the gene-editing technologies with an even greater potential to fundamentally transform our species.

---

*   This is why the advanced space people in so many sci-fi films have such big heads.
†   These people could also become arrogant pricks oppressing us.

Chapter 5

Divine Sparks and Pixie Dust

Revolutionary ideas and revolutionary technologies often advance together.

Experimenting with his pea plants, Gregor Mendel could hardly have imagined the complexity of calculations computers would make possible a century later. Watson, Crick, Franklin, and Wilkins could never have uncovered the double-helix structure of DNA without X-ray photography and the microscope. Fred Sanger, Alan Coulson, Leroy Hood, and others could never have invented genome sequencing without the microprocessor. The army of researchers around the world striving to better understand the human genome would be nowhere without their complex algorithms and advanced processing chips.

This tango between our coevolving tools and ideas is also forever shifting our sense of ourselves as a species. While our ancestors may have seen themselves as the love children of divine sparks and pixie dust, many of us today see ourselves as the output of code. We have given a language to our machines that's become a metaphor for the inner workings of ourselves. It's not just a scientific leap but also a conceptual one. After billions of years of Darwinian evolution by random mutation and natural selection, this shift enables us to envision a future when we will start not only selecting our future children but also hacking and writing their genetic code.

As soon as scientists began unlocking the mysteries of the genome they began imagining how changes could be made. In the 1960s, they started using radiation to spur random genetic mutations in simple organisms and plants, a slow, expensive, imprecise, and painstaking process. For every desired mutation found there could be hundreds, thousands, or even millions of inconsequential or harmful ones. Figuring out in the 1980s and '90s how to more precisely splice genes from one organism to another was a big step forward, but the hunt was still on for a better, faster, and more targeted way of altering genes. More recently, this process has shifted into overdrive.

In an important 2009 study, American geneticists Aron Geurts and Howard Jacob detailed how a class of proteins called *zinc-finger nucleases*, or ZFN, designed to bind to DNA, could be used to precisely edit genomes. With ZFN, proteins are painstakingly engineered to bind and create double-strand breaks to DNA in a specific place. If we imagine the DNA helix as a twisting ladder, ZFN cuts the part where you hold your hands when climbing.

This technology was soon used to edit the genomes of mice, rats, cows, pigs, and other nonhuman mammals in all sorts of laboratory experiments with far greater precision than ever before. ZFN quickly became the predominant gene-editing technology in research labs around the world. Its predominance didn't last long.

In 2011, an even more convenient gene-editing tool was discovered. Transcription activator-like effector nucleases, or TALENs, was yet another obscurely named but revolutionary tool. TALENs also made double cuts to the DNA ladder but was more flexible and versatile than ZFN and could be used to edit a broader range of genetic target sites with greater specificity.

In those ancient times of just a few years ago, TALENs seemed like magic. It was used to model multiple human diseases more efficiently and effectively, and to create genome-edited mice, cattle, pigs, goats, sheep, and even monkeys. As the process improved, it was used to eliminate a genetic eye disease in mice, and its prospects for helping cure human diseases seemed extremely promising. Recognizing its significance, the influential journal *Nature Methods* named TALENs its "Method of the Year 2011."[1] But while TALENs was easier to use than ZFN and seemed relatively superfast at the time, it too was nothing compared to the world-changing gene-editing tool in the making for a quarter of a century or billions of years, depending on how you are counting, but racing around the corner.

This new tool begins with among the smallest of organisms.

X

Bacteria are one of the earliest life forms on earth and the ultimate survivors. Viruses have been attacking bacteria for more than a billion years in their never-ending quest to find hosts into which to insert a tiny package of viral DNA. The viruses aren't doing this out of malice. Hijacking and transforming host cells into tiny virus-producing machines is the virus's sole strategy for survival. The viruses are aggressive, but the bacteria wouldn't have thrived for so long if they didn't build their own defensive survival strategy along the way.

In 1987, researchers from Japan's Osaka University examining a sequence of chromosomal DNA observed a series of genetic code repeats in a series of code clusters. A couple of year later, a young Spanish researcher named Francisco Mojica, who was studying sequenced bacteria with extreme salt tolerance, kept seeing the same types of repeated clusters of palindromic (e.g., madam I'm Adam) code popping up in the bacterial DNA.

When Mojica compared the sequences he'd uncovered with those other researchers had placed in the common GenBank database, he began to notice that the bacteria's palindromic repeated code clusters matched some of the same code clusters in certain viruses. At that time, no one knew what these clusters of code were for or if they were even important. Mojica and other researchers like Giles Vergnaud and Alexander Bolotin in France, however, made a series of brilliant educated guesses that the bacteria were using the code repeats as some type of immune system.[2]

A Dutch researcher, Ruud Jansen, later named these sequences *clustered regularly interspaced short palindromic repeats*. The name was a mouthful, so he shortened it into the more user-friendly acronym CRISPR.

Around the same time, scientists at Danisco, the largest yogurt company in the world, learned about Bolotin's work. Because the Streptococcus thermophilus bacteria is a mainstay in the process of turning milk into yogurt, Danisco scientists Philippe Horvath and Rodolphe Barangou started wondering if understanding how that bacteria responded to a viral attack might suggest new ways of preventing their yogurt and cheese cultures from occasional collapse.

Using the lessons learned from Mojica, Bolotin, and others, Horvarth and Barangou exposed their culture bacteria to viruses, killing most of the bacteria. But when they repeatedly cultured the surviving bacteria and introduced the same viruses, the bacteria got progressively better at fighting them off. The bacteria had, in effect, developed an immunity to the viruses just like we get immunity to chickenpox after an exposure—just as Vergnaud and Bolotin had predicted. Understanding where these CRISPR repeats came from and how they functioned forced researchers to look far back into the deep history of microbial life on earth.

Although the battle between virus and bacteria had been going on for a long time, scientists didn't know much about how it was playing out until new genome-sequencing tools made a deeper level of scrutiny possible. The discovery of CRISPR came at the intersection of genome sequencing and big-data analytics. The "heroes of CRISPR" were code-breakers cracking the genetic code for how the bacteria defended itself.

A CRISPR is like an Old West most-wanted poster of a virus that a bacterium stores in its own genetic code after an initial exposure. The bacteria archives fragments of the viral DNA from these past exposures into the bacteria's own genetic code to create a series of genetic "mug shots" of bad guy viruses.

If the virus rolls into cellular town, the bacteria sends an RNA probe to search for code in the virus's DNA that matches the stored genetic CRISPR target list. When it finds one, the bacteria uses an enzyme to bind to the virus's matching code and cut apart the viral DNA just at the site where the bacterial and viral codes match. When this works, the viral attackers are cut to pieces, the bacteria survive the attack, the piano starts up again, and the saloon customers go back to their card games.

But this story isn't just about viruses and bacteria. Because some bacteria merged into the cells of most other life forms hundreds of millions of years ago, the genetic code that originally stemmed from the ongoing war between viruses and bacteria became replicated in nearly all cells across the spectrum of life. CRISPR, it turned out, held the key to potentially editing the code of all life and changing biological life as we know it.

As more researchers began to explore the science of CRISPR, big strides in understanding this remarkable new tool came fast and furious. In 2010, Sylvain Moineau and his colleagues showed how the CRISPR-Cas9 (for CRISPR-associated gene number nine) system made double-stranded breaks in DNA at predictable and precise locations. The following year, Emmanuelle Charpentier figured out how small bits of two different types of RNA guide the Cas9 enzyme to its target.

Then the following year, in 2012, Charpentier and her new partner, Berkeley biochemist Jennifer Doudna, as well as Martin Jinek, cleverly adapted the CRISPR-Cas9 system into a precise tool that could be harnessed to cut any strand of DNA. They also figured out how the system could be used to insert additional new DNA. After being cut, the DNA tries to rebind where the cut has been made and will grab the available DNA researchers have positioned to fit into the opening. This made the gene-editing process far easier than ever before. The following year, Doudna, Charpentier, and Harvard/MIT researcher Feng Zhang announced that CRISPR-Cas9 could be used to target multiple locations in the human genome at once.

If we had to summarize CRISPR into a single sentence, it would be this: CRISPR systems deploy the same tiny scissors bacteria use to cut up attacking viruses to snip any genetic code in a targeted place and potentially insert new genetic code. The following chart provides another helpful overview.

EDITING A GENE USING THE CRISPR/CAS9 TECHNIQUE

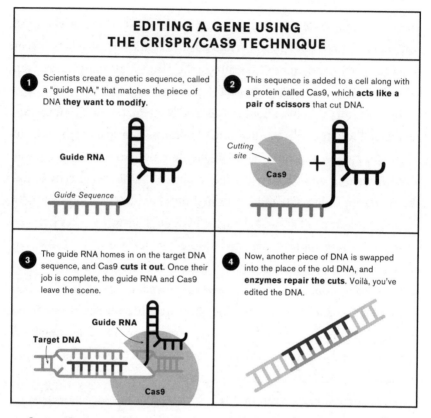

1 Scientists create a genetic sequence, called a "guide RNA," that matches the piece of DNA **they want to modify.**

Guide RNA

Guide Sequence

2 This sequence is added to a cell along with a protein called Cas9, which **acts like a pair of scissors** that cut DNA.

Cutting site

Cas9

3 The guide RNA homes in on the target DNA sequence, and Cas9 **cuts it out.** Once their job is complete, the guide RNA and Cas9 leave the scene.

Guide RNA

Target DNA

Cas9

4 Now, another piece of DNA is swapped into the place of the old DNA, and **enzymes repair the cuts.** Voilà, you've edited the DNA.

Source: *Business Insider.*

The CRISPR-Cas9 system is such a big deal because it has huge advantages over the older gene-editing approaches. While the ZFN and TALENs are bespoke systems, taking months to design, the CRISPR technology is almost entirely always the same, takes less than a few days to set up, and costs relatively little. But despite these overwhelming strengths, the CRISPR system has its shortcomings. Slicing the double strand of DNA using CRISPR was significantly more precise than with ZNF or TALEN, but such an aggressive cut opened the door for even greater unintended effects.[3]*

---

*   An intense debate broke out among scientists after a 2017 study, later debunked and retracted, suggested that the off-target effects of CRISPR were greater than previously understood.

While the scientific world and the popular media were swooning over CRISPR-Cas9, a steady stream of advances was making abundantly clear both that CRISPR-Cas9 was even more versatile than previously understood and that it was not the last word in precision gene editing but just the beginning.

Rather than an equivalent to scissors, the CRISPR system is now looking more like a versatile Swiss Army knife able to record genetic changes within a cell over time, identify specific virus strains, test for infections, spatially reorganize the genome, and perform a host of other functions.[4] Another approach pioneered by the Broad Institute's Feng Zhang pairs CRISPR with a different enzyme, Cas13, to edit the messenger RNA to help better target where cuts and changes to the genome can be made.

Scientists are also now using CRISPR not just to change the genes but also to alter the epigenetic marks dictating how the genes are expressed.[5] Although early skeptics of human gene editing correctly warned that epigenetic influences on the expression of genes made effective gene editing a lot more complicated than first understood, recent advances have made clear that epigenetic editing is "on the verge of reprogramming gene expression at will."[6] As the rate of change in gene editing is accelerating, the process is being made cheaper and more precise, and the globally networked community of scientists is sharing ideas at a level and speed that would have been unthinkable not just to a relatively isolated monk like Gregor Mendel but also to some of the world's most sophisticated scientists only a decade ago.

The step-by-step progression from basic plant and animal experiments and applications to increasingly substantial human uses is already underway.

The first phase of normalization of precision gene editing involves using these tools to advance basic research. One of the most significant early benefits of CRISPR-Cas9 is its ability to target and isolate for study specific DNA sequences. By using the "molecular scissors" of Cas9 enzyme to quickly, easily, and cheaply cut a specific DNA sequence, scientists can study the effect of a compromised gene on cells and organisms. This is a huge deal for scientific research that is increasingly being translated into real-world applications on plants, laboratory and farm animals, and preliminarily humans.

A plant pathologist at Pennsylvania State University, for example, used CRISPR to target a whole family of genes that encodes polyphenol oxidase (PPO), an enzyme responding to oxidization.[7] The gene-edited "Arctic apple" can be sliced and left out in the open without browning because scientists have used CRISPR to silence a gene controlling production of an enzyme that causes apples to brown.[8] A virus-resistant Rainbow papaya gene edited to avoid the devastating papaya ringspot virus is already in supermarkets, as is the bruise-resistant Innate potato. Del Monte has received approval to gene-edit a pink pineapple modified to contain more of the antioxidant lycopene than regular pineapples. A waxy corn that makes better cornstarch, wheat with a higher fiber and lower gluten content, tomatoes better able to grow in warm climates, and camelina with enhanced omega-3 fatty acids are all on the way, thanks to CRISPR technology. In our yogurt and in plants like these, we humans are already bringing CRISPR-edited genes into our bodies.

Gene editing crops isn't just about keeping our mushrooms from browning and our potatoes from bruising. This technology also has the potential to protect billions of dollars of crops and save the lives of millions of the world's poorest people.[9] In Africa and South Asia, where huge percentages of the population are subsistence farmers, average temperatures are warming, and populations are expanding, this need is particularly stark. Using gene editing to create new, more resilient, faster-growing varieties of rice and other crops that need less water could, in the words of Microsoft founder and philanthropist Bill Gates, "be a lifesaver on a massive scale."[10]

Gene-editing tools are also being used aggressively on animals. In addition to generating the gene-edited laboratory rats and mice for biological research, CRISPR is being used to alter the genes and gene expression of a menagerie of other animals. The Gates Foundation, for example, is supporting efforts by the Global Alliance for Livestock Veterinary Medicines to create genetically modified "supercows" that can withstand very hot temperatures while producing much more milk than traditional cows.[11] One scientist whose daughter is allergic to eggs is using CRISPR-Cas9 to engineer hypoallergenic chicken eggs. Researchers are using CRISPR to genetically engineer virus-resistant and faster growing pigs and parasite-resistant and hornless cattle, which could save commercial ranchers hundreds of millions of dollars per year.

On a less practical level, China's BGI-Shenzhen created micropigs designed to be lab animals or pets that grow to a maximum of 33 pounds, one seventeenth of the weight of an average adult pig.[12] Harvard's George Church is even exploring the possibility of using CRISPR and industrial-scale multiplex automated genome engineering tools to revive the extinct woolly mammoth by simultaneously altering multiple genes of Asian elephant embryos.[13]

Wooly mammoths notwithstanding, accommodating gene-edited plants and animals in our food supply and homes will create a greater level of acceptance of gene editing more generally. The incredible promise of gene-editing tools to treat and cure diseases will propel social acceptance of the technologies even more.

Although this book is about the heritable genetic alterations that will ultimately transform our species, the path from here to that future will pass through the application of nonheritable genetic therapies to treat diseases and improve health care. Transforming our evolutionary process is the ultimate destination of the genetics revolution, but medicine will be an essential way station on the road from where we now are to where we are inevitably going.

X

When the possibility of altering a living human's genes to treat disease was introduced in the 1980s, it was recognized as a potential but far-off game-changer. Over the ensuing years, the possibility of manipulating genes to prevent or treat multiple diseases has become an increasingly real possibility. Instead of the more traditional methods of treating diseases with surgery or drugs, gene therapies seek to use genes to treat or prevent disease by knocking out or tuning off a mutated gene, replacing a mutated gene with a healthy copy of the same gene, and/or adding a new gene to help the body fight a particular disease.* As exciting as all this sounds and is, the path toward realizing gene therapies in health care has been a rocky one.

By 1999, doctors at the University of Pennsylvania were confident the science of gene therapy had advanced enough that it could be used to treat Jesse Gelsinger, an eighteen-year-old with ornithine transcarbamylase (OTC) deficiency, a rare genetic disorder that causes increased ammonia levels in the blood often leading to brain damage and premature death. Four days after receiving an infusion of a corrective OTC gene delivered in a manipulated cold virus, Gelsinger died.

Jesse Gelsinger's death was not just a personal and family tragedy, it also put a screeching halt to the application of gene therapies to treat disease in the America, then by far the leading venue for these technologies in the world. The FDA prohibited the University of Pennsylvania from continuing its human gene therapy trials, began investigating sixty-nine other ongoing gene therapy trials in the United States, and started requiring a far higher level of patient safety for gene-therapy trials.[14]

But with the prospect of gene therapy slowed, researchers got to work developing better and safer gene transfer protocols.[15] As gene-editing techniques improved in the ensuing years, so did the prospects for gene therapy. By 2009, the journal *Science* declared the "return of gene therapy" as the major breakthrough of the year,[16] and the more recent development of better gene-editing tools like CRISPR have only made the future of gene therapies seem even brighter.[17]

Although many gene-therapy protocols are now being actively explored, one of the most exciting and widely covered in the media is genetically enhancing the ability of a person's T cells, white blood cells that play an

---

*   The difference between gene therapy and gene editing sometimes confuses people because the two processes are closely related. Genetic engineering and gene editing are tools used in gene therapies that also have far broader functionality. All gene therapy is genetic engineering, but not all genetic engineering is gene therapy.

essential role in the body's natural immune response. In CAR-T therapy, blood cells are extracted from the body of a person with specific cancers and then engineered to boost the ability of their T cells to express a chimeric antigen receptor (CAR) before being put back into the person's body with cancer-fighting superpowers.

In the first three months of clinical trials of this approach by the pharmaceutical company Novartis in 2017, 83 percent of patients showed a significant remission rate in their cancers. Billions of investment dollars are now flowing into scores of companies like Novartis, Gilead, Juno Therapeutics, Celgene, and Servier that are working on gene therapies for cancer, and many hundreds of clinical trials are running around the world. CAR-T therapy still has some significant challenges, and a small number of people have even died during clinical trials, but there is no doubt removing, editing, and reintroducing genes will play a greater role in fighting cancer and other diseases going forward. As of August 2018, the U.S. Food and Drug Administration had received more than seven hundred applications for gene therapy trials.[18]

Removing cells from the body to edit and return them in gene therapy is a massive step forward, but an even bigger advance is our growing ability to use CRISPR and other tools to edit the offending cells still *inside* the body. Editing cells inside the body would open a whole new set of possibilities not only for treating disease but also potentially for gene editing human embryos.

A recent study, for example, gene edited cells to treat diseased human livers grown inside living mice.[19] Another study used CRISPR base editors to precisely correct genetic mutations causing a metabolic liver disease in adult mice.[20] These types of in vivo gene therapies are not yet ready for prime time but is showing promising early results in potentially treating congenital blindness, hemophilia B, beta-thalassemia, Duchenne muscular dystrophy, cystic fibrosis, and spinal muscular atrophy. Incredible new companies like Editas, cofounded by George Church, Feng Zhang, and others, and Caribou Biosciences, cofounded by Jennifer Doudna, and others, are moving forward aggressively to develop new ways of treating multiple diseases with CRISPR gene editing. A forty-four-year-old man in

California in November 2017 became the first person to have genes edited inside his body, in this case to treat the metabolic disease Hunter syndrome.

Responding to this progress, the FDA and National Institutes of Health jointly announced in August 2018 that they would significantly reduce the special oversight processes for gene therapies because "there is no longer sufficient evidence to claim that the risks of gene therapy are entirely unique and unpredictable—or that the field still requires special oversight that falls outside our existing framework for ensuring safety."[21]

This rapid coming of age of gene therapies is being matched by other developments pushing the gene-editing revolution forward. A Silicon Valley start-up named Synthego, for example, sells custom CRISPR-edited human and other cells lines delivered to researchers within days. Another company, Inscripta, is trying to make all of the tools needed for CRISPR just an easy click away. These new companies, *Wired* noted in May 2018, are betting that "biology will be the next great computing platform, DNA will be the code that runs it, and CRISPR will be the programming language."[22] Although the current generation of genetic health-care interventions is not passed on to future generations of the people being treated, the popular acceptance of and demand for these treatments will play an important role making the general public more comfortable with the concept of human genetic alteration.

It would be impossible to capture in this or any book the breadth and speed of the experimental advances with major health-care implications using precision gene editing being made almost daily, but here are a few examples of the intensifying stream of progress made in recent years alone:

- ✗ In 2013, researchers in the Netherlands used CRISPR-Cas9 on human stem cells to repair a defect that contributes to the appearance of cystic fibrosis.[23]
- ✗ In 2014, scientists used CRISPR-Cas9 to correct liver cells in mice modeling the human disease hereditary tyrosinemia.[24]
- ✗ In 2015, researchers deployed CRISPR-Cas9 to edit endogenous beta-globin genes in human cells which, when mutated, result in beta-thalassemia blood disorders.[25]

- x In 2016, scientists used CRISPR-Cas9 to extract HIV from human immune cell DNA and prevent the reinfection of unedited cells.[26]
- x In 2017, researchers first used CRISPR-Cas9 on a human embryo successfully to correct a defect in the MYBPC3 gene that causes hypertrophic cardiomyopathy.[27]
- x In 2018, scientists showed how a novel CRISPR gene-editing technique could potentially correct most of the three thousand mutations causing Duchenne muscular dystrophy by cutting single points along a patient's DNA.[28]
- x In 2019, researchers showed how CRISPR-Cas9 could be combined with special guide RNA to more accurately than ever before edit human cells to correct the genetic mutation causing sickle cell disease.

Because these breakthroughs are happening so rapidly as ideas and innovations cross-fertilize, it is certain that more CRISPR miracles will be announced after the final edits of this book are submitted. It is also certain that new gene-editing tools more precise than CRISPR will arrive in the coming years. "Gene therapy will become a mainstay in treating, and maybe curing, many of our most devastating and intractable diseases," FDA Commissioner Scott Gottlieb presciently declared in 2018.[29]

Getting comfortable with editing people's cells to cure terrible diseases will provide a level of comfort with and confidence in our ability to use CRISPR and other tools to precisely and safely edit the human genome. As this comfort level increases, scientists, doctors, and prospective parents will begin to ask why these tools can't also be used to prevent these diseases in the first place.

X

Mitochondria are the tiny power packs of the cell. Floating in the cell's cytoplasm (if the cell were an egg, the nucleus would be the yolk and the cytoplasm the white), they are the legacy of symbiotic bacteria incorporated into our cells hundreds of millions of years ago. Nearly all our twenty-one thousand or so genes are located in the cell's nucleus, but a far smaller number, just thirty-seven, are in the mitochondria.* Unlike nuclear DNA, which is the combination of the DNA from both parents, mitochondrial DNA (mtDNA) is passed almost entirely from mother to child.

Most people have healthy mitochondria that allow their body to get the energy it needs from their cells. But about one in two hundred people has a disease-causing mtDNA mutation, and about one in sixty-five hundred develop symptoms of mitochondrial disease. These dangerous mutations primarily target children, who often suffer systemic organ failure. The symptoms generally get more severe with age and can increasingly damage cells in the brain, liver, heart, and other bodily systems.

If everyone with mitochondrial disease died young, this disease would have been eradicated from the human gene pool long ago. But a mother's mitochondrial problems tend to be distributed unevenly among her offspring, allowing some of the children to live healthy lives, others to live managing the diseases, and others to die terrible, early deaths.

For millennia, parents with mitochondrial disease had no idea why their children were suffering; they blamed fate. But fate was not a good enough answer for the Swedish endocrinologist Dr. Rolf Luft, who first diagnosed a patient with mitochondrial disease in 1962. Although huge progress was made in understanding mitochondrial disease in the early years, little was achieved in finding a cure or preventing it from being passed mother to child.

In the 1990s, Jacques Cohen and his colleagues at the Institute for Reproductive Medicine and Science in New Jersey pioneered a process of injecting fluid from the cytoplasm of a healthy egg into an egg where problems in the cytoplasm were believed to be causing infertility. Although seventeen babies born through this procedure were mitochondrial disease-free, two fetuses indicated a severe genetic disorder.[30] In response, the FDA in 2001 started requiring clinics to apply for approval to carry out the procedure. Because of the imperfect safety record, few applied. No approvals were granted.

---

*   Some nuclear genes influence mitochondrial function and vice versa.

Nevertheless, the underlying science continued to advance. Over the past decade, teams in the United Kingdom and the United States developed two new mitochondrial transfer procedures. In one, the healthy nucleus taken from an egg of the intended mother carrying faulty mitochondria is removed and placed inside the denucleated egg of a donor woman without mitochondrial disease—it's like keeping the egg yolk but replacing the white with a donated one. In the other procedure, the same process happens to the early-stage embryo after the egg is fertilized; scientists remove the nucleus and place it into the denucleated embryo donor parents provide.

As prospective mothers carrying mitochondrial disease learned of this new approach, many were interested. But some observers were worried. Mitochondrial transfer is a heritable treatment. A daughter born with donor mitochondria will pass that mitochondrial DNA to her daughter, and down the line forever. (This is why women who get their ancestral history through DNA tests can learn about their mother's mother's mother's mother, all the way back to our human female common ancestor, "mitochondrial Eve," from around 160,000 years ago.) Even though the total amount of donor DNA in a child born with the mitochondrial transfer treatment would be small, scientifically altering human DNA for all future generations is a big deal.

The United Kingdom has done more than any other country to thoughtfully consider mitochondrial therapies and their implications. Soon after the "test tube" miracle baby, Louise Brown, was born in Manchester in 1978, the United Kingdom created a Committee of Inquiry into Human Fertilisation and Embryology. The committee produced a major 1984 report on the future of assisted reproduction and then a 1987 White Paper outlining a legislative agenda for going forward. This critically important work culminated in the 1990 Human Fertilisation and Embryology Act that created, you guessed it, the Human Fertilisation and Embryology Authority, HFEA. Since then, the HFEA has done incredible work overseeing and regulating reproductive technologies across Britain.

Although the 1990 act didn't consider, and therefore couldn't expressly authorize, mitochondrial therapies, the issue was raised in 2010 when researchers asked the UK Department of Health and Social Care to amend its regulations to allow mitochondrial transfer. Rather than just considering this a simple regulatory decision, the UK government launched an intensive five-year consultation process, including a series of expert panels, public forums, comment opportunities on draft legislation, and cost-benefit analyses by the department of health. In 2015, the issue of whether the HFEA should be allowed to authorize clinical trials was put to a full vote of both Houses of Parliament and passed unanimously. The HFEA then waited for results from additional studies and convened even more expert panels on the safety and efficacy of these procedures.[31]

In March 2017, Britain's HFEA granted its first license to doctors to use the mitochondrial transfer technique on a human embryo to be implanted in prospective mothers. After the first two clinical applications were approved on February 1, 2018, the first British child born using mitochondrial transfer will all but certainly be delivered in 2019.[32] Moving forward with this first ever case of state-sponsored, heritable genetic engineering was a monumental step not just for the United Kingdom but also for humanity—and the British handled the process responsibly.

In America, the process of considering mitochondrial transfer has been much more bureaucratic. The FDA effectively banned mitochondrial transfer in a 2001 extension of its authorities, and then in 2016 Congress forbade the FDA from even authorizing clinical trials of mitochondrial replacement therapy. Although a series of expert panels on the issue have been held, the FDA has yet to authorize clinical trials partly because the contentious politics of abortion in the United States makes any discussion of manipulating embryos extremely complicated. Even after a 2016 U.S. National Academies of Science, Engineering, and Medicine report concluded that some limited application of mitochondrial transfer treatments could be justified for male embryos (to make sure no genetic changes could be passed on to future generations), the effective U.S. ban on mitochondrial transfer treatments remains in place.

But before the first British license was granted, and while the procedure was still banned in the United States, a Jordanian couple approached
New York–based physician John Zhang about having it done. Because
mitochondrial transfer was still illegal in the United States, Zhang agreed
to travel to Mexico, which at the time had no rules governing the procedure. By the time this birth was officially announced in September 2016,
the Jordanian baby was already a healthy five-month-old. Zhang returned
to New York without repercussion and soon announced the creation of his
new company, aptly named Darwin Life, which he described as "pushing
the boundaries of Assisted Reproductive Technology."[33] In January 2017,
doctors in Ukraine transferred the nucleus of an early-stage embryo with
mitochondrial disease to a denucleated donor embryo without it, resulting
in the birth of another healthy baby.[34]

Potential mothers who carry mitochondrial disease in the few places
where mitochondrial transfer is allowed have a full range of options. Of
course, they can always adopt. But if they want to have a fully biologically
related child, they have the option of rolling the dice by getting pregnant,
testing the embryo after ten weeks of pregnancy, and facing the option of
abortion. Or they can just have the child and see if their naturally born
and unscreened children will be born with a deadly form of mitochondrial
disease. An alternative option would be for the mother to have her eggs
extracted and fertilized using IVF and then genetically screened with PGT.[35]
But mitochondrial disease does not fully show up in early-stage embryos,
so an embryo screened prior to implantation could still be a carrier of
mitochondrial disease.[36]

If a mother in one of these jurisdictions wanted to be certain her child
would not carry mitochondrial disease, she could swap out the cytoplasm
of her egg prior to insemination during IVF, or of her early-stage embryo
just after the egg was fertilized—all for the price of 0.1 percent of the child's
total DNA becoming inherited from the mitochondrial donor.

But what about women living in parts of the world where mitochondrial transfer is banned or unavailable? They can adopt, too. Of course,
they can roll the hereditary dice. They can also do IVF and PGT if they

are in a jurisdiction where that is legal, but they would still run the risk of passing the disease to future generations. They can travel to a place like Ukraine. But that's expensive, uncomfortable, and inconvenient. Another option would be to band together with other prospective parents with the disease to form a lobby group to try to make mitochondrial transfer legal at home.

That's exactly what the mitochondrial disease community has done in the United States. "We strongly support further scientific investigation of oocyte MRT [mitochondrial replacement therapy] as well as constructive debate towards the clinical approval of this therapy in women with mtDNA-related diseases," the Pittsburgh-based United Mitochondrial Disease Foundation, UMDF, publicly declared. "If demonstrated to be safe and efficacious, this technique should be made available as an option to families who carry mtDNA point mutations."[37]

As evidence of the safety and efficacy of mitochondria transfer in places like the United Kingdom continues to grow, pressure on other governments to fund research and ultimately allow heritable mitochondrial therapies by advocacy groups in those countries will increase. Politicians in those countries will have a hard time telling mothers terrified of passing potentially deadly mitochondrial disease to their children and influential single-issue lobby groups that they can't have access to the mitochondrial replacement procedure that has been proven safe and effective under the highest level of government scrutiny in the United Kingdom. Over time, mitochondrial transfer will likely become the first relatively widely accepted heritable genetic manipulation.

Once this happens, parents afraid of passing other deadly genetic diseases to their future children will not sit by idly while their future children face potential genetic death sentences. Instead, they will demand that the most advanced precision gene-editing technologies be used to make the targeted changes that will save their future children from suffering. As scientists continue provide an increasing set of new potential possibilities to gene edit embryos to prevent disease and enhance health, parental demand will only increase.

Almost every significant genetic disease has its own social network, and many also have politically influential lobby groups. Disease advocacy groups spend many millions of dollars annually lobbying the U.S. government. Each thousand dollars invested in this type of outreach is estimated to correlate to a $25,000 increase the following year in National Institutes of Health funding for a specific disease.[38] It is hard to imagine the U.S. government, itself driven largely by the interests of special-interest pressure groups, not eventually supporting research into and clinical trials of the most promising treatments for genetic diseases, even those that involve making heritable changes to preimplanted embryos.

The major strides toward making gene editing of preimplanted human embryos possible will add more fuel to this rocket.

In April 2015, scientists from Sun Yat-sen University in Guangzhou, China, shocked the world by disclosing they had used CRISPR-Cas9 to genetically alter genes in human embryos in vitro linked to the often-fatal blood disorder beta-thalassemia.[39] The embryos were nonviable because they had been fertilized by two sperm cells instead of the usual one, and the accuracy rate of the edits was dismal. But this first reported direct application of CRISPR to the nuclear DNA of human embryos crossed an ethical Rubicon in the minds of many observers.

Soon after, United Kingdom regulators approved an application by Kathy Niakan. A researcher at London's Francis Crick Institute, Niakan wanted to CRISPR-edit the genes of viable human embryos and examine how a gene called *OCT4* regulates the development of fetuses, a first step to better understand a particular cause of infertility. Two months later, another team of Chinese scientists announced they had used CRISPR to try to make early-stage, unimplanted human embryos resistant to HIV.[40]

Then, in July 2017, the path-breaking and controversial U.S. scientist Shoukhrat Mitalipov of Oregon Health and Science University became the first American researcher to use CRISPR-Cas9 to genetically alter human sex cells and unimplanted embryos. Mitalipov injected a CRISPR-Cas9 genetic scissors into the sperm of a man carrying a faulty MYBPC3 gene, which can cause hypertrophic cardiomyopathy, a heritable disease that can lead to sudden heart failure in children. When this gene-edited sperm was used to fertilize the eggs of twelve healthy donor women, two-thirds of the embryos were created disease free—a big increase over previous efforts. When Mitalipov's team tried the same thing but only injected the unedited sperm and the CRISPR-Cas9 separately—so that the sperm editing happened simultaneous to egg fertilization—the success rate increased to 72 percent. Seventy-two percent efficacy is still not nearly good enough, and the embryos were all destroyed within three days, but a marker had clearly been set down on the path to the heritable gene editing of nuclear DNA in humans.[41]

"We've always said in the past gene editing shouldn't be done, mostly because it couldn't be done safely," MIT researcher Richard Hynes told the *New York Times* after Mitalipov's findings were released. "That's still true, but now it looks like it's going to be done safely soon."[42] When this happens, the irresistible attraction of using our most advanced technologies to eliminate our most deadly diseases will draw us into the genetic age.

The amazingly rapid transfer of advanced gene-editing tools from laboratories to farms and ranches and then to our hospitals and fertility clinics is already well underway with a momentum of its own. Almost every day, a new application is announced that, once available, some group of humans will demand. The very real benefits of these technologies to a growing group of potential beneficiaries will eventually outweigh the abstract aspirations of a dwindling number of genetic traditionalists.

But as we increasingly master and grow comfortable with this technology, we'll create a process with applications extending far beyond health care, which we'll use to alter the genetics of ourselves and our children in increasingly significant ways.

Chapter 6

Rebuilding the Living World

Death from exposure to the Ebola virus is excruciating.

First, you feel extremely weak, with flu-like symptoms. As the virus starts infecting and then bursting your cells and blood vessels, you then experience uncontrolled nausea, diarrhea, vomiting, and headaches. Your cells hemorrhage, causing uncontrolled bleeding throughout your body. You then go into shock before you die a gruesome, bloody death, your vital fluids bursting out from every orifice of your body.

Early Ebola outbreaks in the poorest parts of Africa saw death rates of up to 90 percent of those infected. The somewhat better health care provided during the 2014 Ebola outbreak in West Africa reduced the death rate to around 60 percent.

The people most likely to be infected by Ebola are the family members and health-care providers caring for loved ones and patients already infected. Just being exposed to the saliva, vomit, urine, or stool of an Ebola victim is enough.

But scientists who studied survivors of the 2014 Ebola outbreak in Guinea were surprised to come across a group of caretaker women who had been exposed but somehow seemed immune to the disease. Some of these women had contracted the virus at an earlier point and survived, possibly giving them immunity to later exposures. Others, however, possessed antibodies even though they hadn't been exposed. *What was happening?* the scientists wondered. Could some of these women be genetically immune to Ebola?

Researchers zeroed in on a gene that encodes a protein called *Niemann-Pick Type C*, or NPC, which the Ebola virus targets when attacking a host. Having nothing to do with Ebola, children inheriting two copies from their parents of the mutated version of this gene generally die from Niemann Pick Type C disease, a neurodegenerative disorder.

Some single-gene diseases like Huntington's disease and Marfan syndrome are dominant, which means that you'll almost certainly get the disease if you inherit the gene and either of your parents is homozygous for the mutation. Others, like with sickle cell disease, Tay-Sachs, and Niemann-Pick Type C disease, are recessive disorders, which means you only get the disease if you inherit the gene from both parents. But just like recessive carriers of the sickle cell disease gene can have immunity from malaria, preliminary studies of these West African women suggested that people with only one mutated copy of the NPC gene may have increased resistance to the Ebola virus.

We've already explored single-gene and other mutations that can cause disease. But just as there are some individual genetic mutations that can inflict harm, there are many more single-gene mutations that can do good.

As the Ebola case shows, sometimes these can be the same genes that help us in one context but hurt us in another. Over the past decade, scientists have been searching to find more of the single gene mutations with outsize potentials to help us, at least in the context of the world we know today. Finding them, however, is a challenge.

It's much easier to identify an illness that presents itself with observable symptoms than to identify the absence of a disease among people who, but for a particular genetic mutation, might otherwise have it. But by finding outliers like the women with Ebola immunity and pouring through databases of hundreds of thousands, or even millions, of genomes and health records to find correlations between the presence of a rare variant gene and resistance to specific diseases, researchers are increasingly striking gold.

David Altschuler, for example, while a researcher at the Broad Institute of Harvard and MIT, recruited a cohort of elderly and overweight people who were statistically at a high risk of developing type 2 diabetes but hadn't. After sequencing the members of the group, to see how they might be genetically different from other people with the disease who were equally old and overweight, he came to believe that a single mutation in the SLC30A8 gene made his cohort 65 percent better able to regulate their insulin levels and less likely to get diabetes.[1]

Another study found that 1 percent of Northern Europeans carried a mutation in their CCR-5 gene that makes them immune to HIV infection.[2] Another elderly study found that about one in 650 people with mutations of their NPC1L1 gene had less than half the risk of a heart attack compared to people just like them but without the mutation.[3]

Because making small changes to the human genome is easier and safer than making big ones, identifying single-gene mutations with significant potential positive impact raises the enticing possibility of gene editing these small mutations into ourselves or future children. Because our biology represents a delicate balancing act of priorities played out over billions of years, only a tiny number of genes are likely to have an impact large enough to make the benefit of adding or removing them outweigh the potential danger of making a change.

But as the Ebola and other cases point out, it's definitely worthwhile to look for them.

According to Harvard's George Church, a preliminary list of these rare single genes that could potentially be manipulated to give us special benefits might include:

GENE	IMPACT
LRP5	Extra-strong bones
MSTN	Lean muscles
SCN9A	Insensitivity to pain
ABCC11	Low odor production
CCR5, FUT2	Virus resistance
PCSK9	Low coronary disease
APP	Low Alzheimer's
GHR, GH	Low cancer
SLC30A8	Low type 2 diabetes
IFIH1	Low type 1 diabetes

Source: "A Conversation with George Church on Genomics & Germline Human Genetic Modification," *The Niche Knoepfler Lab Stem Cell Blog*, March 9, 2015, https://ipscell.com/2015/03/georgechurchinterview/.

Making small changes to genes like these isn't our only path to effect these changes. Genes, as we've learned, are a set of instructions telling cells to make proteins that do things. Although it matters what the genes say, what the cells actually do is ultimately what's important. So, even if we find a particular gene that does something we feel is good or bad, we wouldn't necessarily need to change the gene in order to change its expression. In some cases, it might make more sense and be safer, easier, and cheaper to develop drugs that instruct the cells to do what we want, even if the "bad" mutation remains or the "good" mutation is never there.

Even so, there will be some mutations, both cumulatively helpful and hurtful ones, where this type of treatment will not be possible. Like with mitochondrial disease, there might be a mutation that is seen as so harmful that some carriers would want to get rid of it for all future generations. Some parents might want to alter the germ line of their future children to make them immune to HIV, less likely to face cognitive decline as they age, or benefit from any other single-gene mutation that confirms a special advantage.

X

When considering the possibility of having one or a small number of single-gene edits made to their preimplanted embryos, the first question parents will ask is whether this is safe. At present, the answer is no. CRISPR is still not a perfect technology. One of the largest concerns about the first generation of CRISPR gene-editing tools was their potential to cut the genome in places other than where the scientists intended.

This type of off-target cutting has shown up most significantly in the gene editing of human cells. An important 2013 study examined off-target CRISPR edits in human cells and found that CRISPR edits for therapeutic applications would need to be significantly improved to be "used safely in the longer term for treatment of human diseases."[4] If these types of off-target mutations were always benign, any small change made by CRISPR editing human genes inside a person wouldn't amount to much. But that is not the case. A CRISPR-induced mutation could also have the potential to become cancerous. That's why regulators around the world have been justifiably cautious about authorizing the gene editing of humans.

The Chinese research group that shocked the world by announcing its CRISPR-editing of the nonviable in human embryos in 2015, for example, reported abysmal accuracy levels. Of the eighty-six fertilized eggs injected with the CRISPR-Cas9 system designed to edit their genomes, only a few contained the desired genetic change. This ratio of attempts to successes is acceptable in plant, worm, fly, and mouse models, where the cost of mistakes tends to be lower but would be unthinkable for humans.[5]

The path toward reaching a level of reliability, where CRISPR could be safely used inside humans, will not be linear. A high-profile 2018 study, for example, found that a single human gene, p53, blocked CRISPR edits in some human cells, as part of the body's natural defense mechanism against dangerous mutations like cancer.[6] One way around this would be to deactivate the p53 gene, but this would bring a new danger—increased cancer risk. Another 2018 study published in *Nature Biotechnology* found 20 percent more off-target DNA alterations than previously expected when making CRISPR-Cas9 edits in mice.[7]

Scientists intent on addressing such concerns have focused on increasing the accuracy of CRISPR gene editing, with some dramatic success. They are discovering new enzymes that more precisely attach to or break the genome than CRISPR-Cas9. These new CRISPRs—with names like CRISPR-cpf1 (a.k.a. 12a), CRISPR-Cas3, CRISPR-13, CRISPR-CasX, and CRISPR-CasY—are proliferating. New AI algorithms are also being deployed to assess where CRISPR edits can most optimally be made.

In 2017, researchers reported a new method for changing DNA and RNA nucleotide "letters"—the A's, C's, G's, and T's—without cutting the genome.[8] The original CRISPR model required cutting across the twisting ladder of DNA; the modified process changes the genes without cutting the ladder at all.

To do this, researchers tricked the DNA atoms to pair differently than they otherwise would. Remember that A's pair with T's and C's with G's; so, if the cell thinks an A is a C, for example, it will pair it with a G rather than a T. The gene and its expression change, but the uncertainty that arises from cutting DNA is avoided.

This approach is particularly useful because an estimated 32,000 of the roughly 50,000 known changes in the human genome associated with diseases are caused by the swapping, deletion, or insertion of a single gene.[9] Called an *adenine base editor*, or ABE, this new version of CRISPR works 34 to 68 percent of the time, with less than 0.1 percent of cells showing evidence of additional mistakes, a big improvement but still not ready to be deployed inside a human body.[10] Chinese researchers reported in August 2018 that they had base edited the genomes to repair a mutation causing Marfan syndrome in sixteen out of eighteen viable preimplanted human embryos.[11] Although none of these embryos were implanted due to legal and ethical considerations, it is clear where this technology is headed.

Base-editing technology was then used to increase the precision and potentially the safety of gene editing even further through a process called CRISPR-SKIP. Editing a single base with this approach causes the cell to "skip over" and not "read" targeted strings of protein-coding genes. Preliminary indications suggest CRISPR-SKIP could be used to deactivate damage-causing mutations in the genome with far fewer off-target effects than many of the other CRISPR systems.[12]

In addition to using CRISPR for gene editing, significant progress is also being made using CRISPR to edit the epigenetic marks orchestrating how genes function and the RNA guides that translate the genetic information into instructions for the cell.[13] Together, these approaches will make altering genes and their expression more precise.

Another big challenge to overcome to make the gene editing of preimplanted human embryos safe is the uneven spread of a genetic change across the cells, which scientists call *mosaicism*. Uneven distribution of a gene-edited mutation can lead to abnormal growth of the fetus and other serious problems. But this challenge, too, is being gradually addressed. Recent studies have shown that using CRISPR as soon as possible after fertilization and editing the sperm and egg cells prior to fertilization decrease the chances that cell mosaicism will occur.[14]

After Shoukhrat Mitalipov and his team announced their new approaches for minimizing this potential problem,[15] another group of superstar geneticists issued a statement the following month raising questions about the accuracy of this research. "It is essential that conclusions regarding the ability to correct a mutation in human embryos be fully supported," Dieter Egli, George Church, and their colleagues wrote in a joint statement arguing that Mitalipov's conclusions were far from proven. "Absent such data, the biomedical community and, critically, patients with disease-causing mutations interested in such research must be made aware that numerous challenges in gene correction remain."[16]

This debate among leading genetics researchers then ramped up a notch in August 2018, when *Nature* published, in the same issue, two scathing critiques of Mitalipov's research as well as a long and detailed response from Mitalipov and thirty-one of his colleagues from around the world.[17] Although all scientists would agree that our ability to precisely edit preimplanted human embryos is increasing, the debate is still raging about whether we are ready yet to use these technologies on human embryos intended to be implanted in a mother and taken to term.

But, once again, the operative word in this last sentence is *yet*.

If logic were our guide, people would start getting more comfortable with gene editing embryos once the error rate in gene editing matched the error rate of natural conception. As we've seen from the experience of self-driving cars, however, the reality is that a new technology like this needs to be much safer than nature for it to be adopted. At least for the technical process of making a very limited number of genome edits, this standard will be soon reached. If and when this happens, gene editing preimplanted embryos and eggs and/or sperm may be the only way for some parents carrying a subset of genetic disorders to have a biologically related child who would not inherit the disorder. These instances would include some Y chromosome defects, dominant monogenetic diseases like Huntington's where one parent is homozygous, and recessive conditions where both parents are homozygous.[18]

An extremely controversial first use of CRISPR to allegedly edit a single gene, CCR5, in the preimplanted embryos of a pair of twins to make them immune to HIV was announced by Chinese researchers in late November 2018. Although roundly condemned by many scientists and ethicists in China and around the world, this first-ever case of gene editing humans was a harbinger of where our genetically engineered future is heading.[19]

But while increasing confidence with the use of IVF, embryo selection, and single-gene editing of preimplanted embryos seems all but inevitable, the prospect of editing more complex genetic traits remains significantly more remote.

As we've seen, complex traits like height, intelligence, and personality are most often determined by the complex interaction of hundreds or even thousands of genes, all performing multiple functions and interacting with other body systems and the constantly changing environment around us.[20] A group of Stanford researchers recently argued that most genetic diseases and traits are not just polygenic, influenced by multiple genes, but instead what they call *omnigenic*. This hypothesis argues that traits are influenced not just by the systemic contribution of the many "core genes," the ones that now show up on genome-wide association studies, but also by a much larger network of peripheral genes that don't.[21] If true, this would make understanding complex diseases and traits far more complicated than previously imagined.

The more genes that influence a particular trait, the more difficult a computational task it becomes to fully understand the correlation between genetic patterns and certain gene expressions. The harder it is to understand the multiple functions each gene performs in the complex and interconnected ecosystem of the genome, the tougher it becomes to make bigger gene edits intending to influence complex traits without unintentionally damaging the rest of the genome.

It certainly should be our assumption that the interconnected ecosystems of the human body are almost always more complex than we tend to think they are. It also makes logical sense that our diseases and traits span a wide range of genetic foundations, from the single-gene mutation diseases like Huntington's and single-gene-determined traits like having wet ear wax on the one hand to complex diseases and traits like coronary heart disease and personality style on the other.[22] The omnigenic model may be a worst-case scenario for aspiring genetic engineers that, even if true in some instances, certainly won't apply equally to all diseases and traits.

But we won't need anything even approaching an omnigenic level of understanding when people are selecting from among their own natural and unedited preimplanted embryos during IVF and PGT. The steady increase in our understanding of complex genetic patterns, even omnigenic ones, will be enough to inform moderately educated guesses by parents about which embryos to implant in the mother. As our imperfect knowledge of these complex traits increases, so will our confidence in selecting and ultimately genetically altering our future children.

X

To many religious people and others who believe in concepts of the spirit and soul, a human being is infinitely complex. For these people, even the most accurate medical tests cannot unlock the mysteries of the spiritual world or the deep interaction between humans and the divine. For those like me, who believe we evolved from microbes, humans are single-cell organisms gone wild over six hundred million years of random mutation and natural selection. We are not infinitely complex beings, just massively complex ones. There's a big difference. If we are infinitely complex, we'll never understand ourselves. If we're only massively complex, there will come a time when the sophistication of our tools will outmatch our own complexity.

We can understand simple organisms pretty well today because our advanced tools are increasingly a match for the complexity of their biology. Human biology, however, remains significantly more complex than our understanding of, or tools to manipulate, it. This won't always be the case. As our knowledge and tools progress, our complexity will become as understandable with our tools of tomorrow as simple organisms have already become with our tools of today.

To get a glimpse of how it will feel when the sophistication of our knowledge and tools begins to surpass the complexity of our biology and what that might mean, we only need to look at how quickly our understanding of single-cell and other simple organisms has grown.

The *C. elegans* roundworm is the perfect example. Smaller at maturity than the size of a comma on this page, the standard mature *C. elegans* goes from birth to death in about two weeks, has a rudimentary nervous system with a brain, reproduces exponentially, and is probably the simplest living organism that shares many of its genes with humans. These qualities, along with its transparent and easily observable body, make these little creatures perfect research subjects and among the most studied model organisms in science.

Over past decades, researchers have starved, chilled, and heated *C. elegans* as they look for the outliers; they have probed the roundworms with advanced microscopes, spun them around wildly in centrifuges, exposed them to antibodies, tagged them with laser microbeams to eliminate individual cells, profiled their proteins, and isolated and amplified their individual genes for closer examination.

In 2011, a group of ambitious scientists came together to create the OpenWorm project, designed to network *C. elegans* researchers from around the world in a common effort to crack the code of how the *C. elegans* functions.[23]

The *C. elegans* has exactly 302 brain cells, compared to about 100 billion for us (unless you did drugs in college or drink diet soda), that have been mapped in a connectome, a diagram of the switchboard of connections showing the worm's brain at work. As an early step toward completely simulating the *C. elegans* as a virtual entity, the OpenWorm collaborators translated its neurons into a computer program used to animate a small robot, where the worm's motor neurons, nose, and other body parts all had robotic equivalents.[24] When booted up, the robot's movements tracked those of an actual roundworm with surprising consistency.

The OpenWorm Robot. Source: "Worm Robot Sneak Peek," OpenWorm, YouTube, published July 21, 2017, https://www.youtube.com/watch?v=1wj9nJZKIDk.

We have moved from little to deep understanding of how the roundworms function in the past few decades because the growing sophistication of our tools has enabled the growing sophistication of our knowledge.

Getting from where we are now in our understanding of our own biology to the point where we will be the equivalent of the *C. elegans* relative to the sophistication of our knowledge and tools will require a map that is just starting to be built. The Human Cell Atlas is an open "coordination platform" that integrates data on human biology from around the world. This collection of "comprehensive reference maps of all human cells" will grow over time as research tools become more powerful and as researchers' knowledge compounds.[25] In these early days, the magnitude of the body's complexity will tower over the capacity of our still relatively humble tools and our current limited knowledge. But why this will change can most clearly be explained in the graph on the next page.

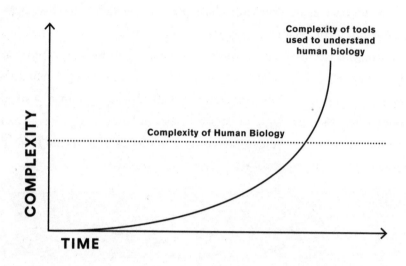

Our biology is about as complex as it has been for millions of years, but the sophistication and capacity of our tools is now advancing at exponential rates.

The basic idea of exponential change is that innovation begets innovation. The more and better tools we develop, the more efficiently and effectively we organize ourselves and connect with each other, and the more ideas we have, the better positioned we are to devise even better tools, organize and connect more, and come up with even greater ideas. That's why it took around twelve thousand years to go from the agrarian revolution to the industrial revolution but only a couple of hundred years to move from the industrial revolution to the internet revolution. Each technological revolution enables the next, and the time between each revolution becomes shorter and the impact bigger. Futurist Ray Kurzweil called this process the "Law of Accelerating Returns."[26]

In 1999, Kurzweil predicted that the entire amount of technological change achieved in the twentieth century would be achieved in only the first fourteen years of the twenty-first century. As innovations compound, he suggested the full twentieth-century amount of change will take only seven years starting in 2021. The same twentieth-century amount of change will take around a year a short while after that time. Ultimately, the equivalent of the full twentieth-century rate of change, equivalent to going from horse and buggy to the international space station, will take only months.[27]

Since the invention of the microprocessor in the early 1970s, this acceleration has been fed by Moore's law, the observation that computing power roughly doubles about every two years for the same cost, a trend that has continued for nearly half a century. It's because of Moore's law that we now expect our smartphones to be lighter and faster with each version, but that's the least of it.

The internet revolution has given us virtually unlimited access to information and to each other, and a network of thousands of educated people linked and working together to solve the same series of problems is way more than a thousand times more creative than the sum total of each of us working alone. Collaborating with artificial intelligence (AI) agents, this networked group of people has the potential to be many times more innovative than the thousand people merely connected to each other. As AI capacities grow, a future form of superintelligence may become more sophisticated than all humans put together, and new possibilities for remaking our species and our world will emerge. Over time, hard things will become easy, and today's complex things will become, relative to the tools and capacities we have to understand them, simple.

X

Collectively as a species, we are today moving along the spectrum, from laying the technological foundation for human genetic engineering to finding preliminary applications to imagining what might be possible in the future to making that imagined future real. Even the legends of our past are strangely becoming our new realities. The idea of humans merging with animals, one of our most ancient human mythologies, is a perfect case in point.

The word *chimera* comes from a Greek word meaning *she-goat*. In Greek mythology, the Chimera was a hybrid of different animals, usually a lion with a goat head and sometimes with a snake tail. In the *Iliad*, Homer described a "thing of immortal make, not human, lion-fronted and snake behind, a goat in the middle."[28] In Dante's *Inferno*, Geryon represents a medieval version of the Chimera, "the monster with the pointed tail... The face was as the face of a just man...and of a serpent all the trunk beside."[29] Many ancient cultures shared this idea. The Chinese *Qilin*, a mythical creature with the neck of a giraffe, the antlers of a buck, and the scales of a fish, allegedly marks the birth and death of important rulers. The Indian *Ganesh* is the son of the gods Shiva and Parvati, with the head of an elephant and the body of a man.

More recently, chimeras have come to mean any creature that is composed of parts from multiple plants or animals, a concept that has moved from fiction to fact.

Nearly a hundred years ago, insulin removed from cows was first used to treat diabetes. Extracted dog, pig, and cow insulin made it possible for humans to live with diabetes and saved countless lives for decades until E. coli bacteria could be genetically modified to produce human insulin at scale.

Over three decades ago, doctors started using pig and cow valves when repairing human hearts. Although this process was somewhat controversial among some Jews, Muslims, and Hindus,[30] the use of decellularized animal aortic valves has become a mainstay of cardiovascular transplant surgery, even for many Jews, Muslims, and Hindus. The benefits simply outweigh the risks.

Scientists have had greater success using animal heart valves for human transplantation than full-organ, animal-to-human transplants. In 1984, California surgeon Leonard Bailey and his team famously transplanted a baboon heart into Baby Fae, a child born with hypoplastic left heart syndrome, a rare congenital heart defect. Although a small number of additional animal-to-human transplants have been attempted since, all of them proved unsustainable after being rejected by each person's immune system.

But although Baby Fae tragically died within a month, the procedure opened the door to progress in human-to-human organ transplants, which have proven far more sustainable than animal-to-human transplants, that saved hundreds of thousands of lives since. These transplants, however, have two major problems. First, the human body is fine-tuned to reject alien DNA, so people getting transplants need to take immunosuppressant drugs for the rest of their lives, which puts them at risk for other ailments. Second, the United States and many other countries face a tragic shortage of donated organs.

As of August 2017, there were more than 114,000 people waiting on organ-transplant lists across the United States. Twenty people die every day while waiting for a transplant. This chronic and deadly organ short-age should, theoretically at least, be solvable with policy changes alone.[31] A single donor can donate as many as eight organs, meaning that he or she could potentially save up to eight lives. But that's not what's happening. In the United States, 95 percent of people support organ donation, but only 54 percent are registered donors.[32] A far smaller percentage actually ends up donating because family members can be conflicted about organ donation at the emotionally challenging moments when the decisions are made.

Because the prospects for getting more humans to donate their organs are still poor in the United States and many parts of the world, scientists have been exploring how new technologies like CRISPR-Cas9 can be used to make animal to human organ transplants more feasible.

Of all domesticated animals, pigs are the most important potential source for transplants, because we already breed a lot of them and because their organs are similar in size and function to ours. But harvesting organs from pigs has at least two big problems. The first, the human body's natural immune response, is extremely serious but has been made more manageable over past decades with both the steady advance of new immunosuppressant drugs and by preliminary efforts to edit the animal genes most likely to be rejected by humans.[33] The second problem, the risk of pig and other animal viruses infecting humans after transplants, is also starting to be overcome.

Pigs carry active viruses called *porcine endogenous retrovirus*, which scientists have given the unfortunate acronym PERV. These viruses can be extremely dangerous and even deadly to humans, particularly those whose immune systems are already suppressed by drugs. Until recently, PERV was essentially a deal-killer for pig-to-human transplants.

Now, however, a group of Harvard scientists is using CRISPR-Cas9 to edit the genome of pig embryos in multiple locations simultaneously to create pigs with their PERVs rendered inactive.[34] Clinical trials for transplanting gene-edited pig kidneys and pancreases to humans are likely to begin soon, potentially saving thousands of human lives per year.

But why stop there? If using insulin made from genetically engineered bacteria and yeast was a step up from using animal insulin, wouldn't transplanting an entire human organ grown outside a person's body from a person's own cells be better than transplanting a gene-edited animal one? What if we could *grow* human organs inside animals for transplantation? Making this possible would be no easy feat, but a significant first step has been made to produce chimeric embryos generated from cells of one type of animal grown in another.

Scientists at San Diego's Salk Institute injected different levels of human stem cells into fifteen hundred pig embryos until they finally found a human cell that integrated into the pig cells. Another team inserted mouse pancreatic cells into rat embryos that grew into mouse pancreases inside the rats, which were then successfully transplanted back into mice to treat diabetes. This work was then followed up by an announcement in February 2018 that a team from the University of California, Davis, had developed sheep embryos with 0.001 percent human cells.[35] Just like with the rats and mice, hacking the sheep embryo to shut off its ability to grow a particular sheep organ and replace it with inserted human genes instructing it to instead make a human version of the same organ type is starting to seem like a real possibility.

Growing human organs in other species will not happen tomorrow, but if it becomes possible for an individual person to have an organ grown inside an animal using his or her own genetics, people needing replacement parts because of illness or the ravages of age will quickly overcome any squeamishness they may have about crossing the human–animal barrier. Governments will be hard-pressed to prevent people from having replacement parts grown in animals.

The possibility of inserting animal DNA into the human genome imagines a future that science is fast making real. The science already exists to relatively easily place a single fluorescent protein from jellyfish into humans that can make a person glow under UV light. If scientists find a single gene or two that make naked mole rats completely resistant to cancer, for example, might we want to splice versions of that gene into humans using CRISPR or some future gene-editing tool? Integrating whole genetic systems like those that give dogs special hearing abilities, eagles amazing vision, or dolphins sonar would be a lot more complicated and not possible any time soon. But the transition of biology into yet another domain for human engineering will, over time, blur our sense of where science fiction ends and science begins.

The possibilities for this type of engineering are vast because all life operates according to different manifestations of the same genetic components. It will someday be possible to even create new human traits and capabilities from scratch, using new combinations of these same genetic building blocks.

X

The exploding field of synthetic biology uses computers and laboratory chemicals to write new genetic code that nature never imagined and make organisms do things they weren't previously programmed to do. Some of its early applications include current efforts to culture meat in a lab, engineer oil-secreting bacteria, manufacture yeast with spider DNA to make ultra-light silk stronger than steel, or induce bovine collagen to make nonanimal leather. Synthetic biology is being used to create renewable microbes to produce acrylics for paints and custom engineer inexpensive synthetic sugars for biofuel. The preliminary list of these types of synthetic biology applications is almost endless. This science and the industry associated with it are exploding as synthetic biology tools become more easily accessible.

These new life forms are being created from readily accessible genetic parts.

The International Genetically Engineered Machine (iGEM) Foundation, for example, provides a free collection of DNA sequences encoding for a particular biological function that can be "mixed and matched to build synthetic biology devices and systems."[36] The BioBricks Foundation, established by MIT and Harvard researchers, provides synthetic gene sequences for free,[37] making ordering genetic sequences about as easy for a researcher as dropping by Home Depot is for a builder. The ease, accessibility, and flexibility of these tools are empowering a synthetic biology revolution where all sorts of useful products, including computer chips, appliances, and clothes, could be grown through engineering biology.

The commercial implications of this revolution are massive. It is estimated that the global synthetic-biology market will grow from around $3 billion in 2013 to about $40 billion in 2020, with a projected global compound annual growth rate of 20 percent. Not surprisingly, China is expected to be the world's fastest-growing market for synthetic biology products.[38] According to leading biologist Richard Kitney, synthetic biology has "all the potential to produce a new major industrial revolution."[39]

As the global population rises, our climate continues to heat, and new and unforeseen challenges arise, these types of precision gene-editing-based synthetic biology applications will become essential for our survival. Our growing dependence on synthetic biology in our lives will pave the way for greater acceptance of synthetic biology in ourselves. This process, too, has already begun.

In 2010, the maverick scientific entrepreneur Craig Venter announced that he and his colleagues had synthesized the full genome of a bacteria called *Mycoplasma mycoides* and placed it into the empty membrane of another bacteria—creating the world's first synthetic cell.[40] This wasn't just editing an existing cell, like getting a bacterium to produce insulin, but building life from scratch. For people concerned that biologists were "playing god," this was Exhibit A. Six years later, Venter's team announced they had pared down the genetic code of their synthetic cell to a much smaller number of essential genes required to keep it alive. As a first step in a never-ending process of creating and rewriting the code of life, this was a major achievement.

"Recent leaps in the biosciences, combined with big data analysis, have led us to the cusp of a revolution in medicine," Venter wrote in a December 2017 *Washington Post* editorial. "Not only have we learned to read and write the genetic code; now we can put it in digital form and translate it back into synthesized life. In theory, that gives our species control over biological design. We can write DNA software, boot it up to a computer converter and create unlimited variations of the gene sequences of biological life."

It is a very long way from synthesizing the genome of a single-cell organism, which hasn't yet fully happened, to doing it for the twenty-one thousand or so protein-coding genes in the human genome but, as the Chinese proverb goes, a journey of a thousand miles begins with a single step. As Stanford synthetic biologist Drew Endy told NeoLife in 2018, "We want to make modellable biological systems that we understand and can use as engineers to rebuild the living world... I have to imagine that one day we'll be routinely building human genomes for any purpose that anyone can advocate for."[41]

Mommy? A DNA synthesizer machine. Source: Seth Kroll.

As computing power increases and the cost of producing an entire human genome decreases, the walking pace will speed up. Lao-Tzu, the purported originator of that Chinese proverb twenty-five hundred years ago, would have needed about three hundred hours to walk that thousand miles. Today, Lao-Tzu's descendants can go a thousand miles in twelve hours by car, four hours by train, two hours by plane, and three and a half minutes in an orbiting spaceship. A journey of a thousand miles still begins with a single step, but travel speeds can accelerate pretty quickly from there.

The Genome Project-write (GP-write), one notable example of this accelerating change, seeks to raise $100 million to synthesize the full human genome, starting with synthesizing the genomes of simpler organisms but continually moving up the complexity chain toward creating human genetic code.[42] "What we're planning to do is far beyond CRISPR," George Church said at the time. "It's the difference between editing a book and writing one."[43]

If funded and even minimally successful, this initiative will help scientists better understand the complex genetic and broader systems biology ecosystem. In the longer term, this understanding will help future generations manipulate, engineer, and ultimately create life. The transformation of human life into information technology will proceed apace: reading, writing, and hacking. Given these aspirations, it's no coincidence that Church's 2012 book is titled *Regenesis*. Our religious traditions have a three-letter word for the entity that writes the book of life.

The computing, machine learning, AI, nanotechnology, biotechnology, and genetics revolutions all have different names today, but these different technologies are currents converging into one megarevolutionary tidal wave, washing over what it means to be a human being. If we ride this wave, the only limit for how far we can go is, perhaps, our collective imagination.

Geneticist Christopher Mason of Weill Cornell Medical College is already, for example, working on what he calls a "ten-phase, 500-year plan for the survival of the human species on Earth, in space, and on other planets" and collaborating with NASA to "build integrated molecular portraits of genomes, epigenomes, transcriptomes, and metagenomes for astronauts, which help establish molecular foundations and genetic defenses for long-term human space travel."[44] The spaceship of our genetic future is already loading at the dock.

That is why when we traveled in our time machine described at the start of this book, the baby we brought back to today from a thousand years ago was pretty much like us, but the baby we brought back from a thousand years in the future was healthier, stronger, smarter, and more robust than most of us are today. It's also why that baby, if we feed and nurture her, will also very likely live significantly longer than any of us do today.

Chapter 7

Stealing Immortality from the Gods

In the world's oldest surviving literary work, King Gilgamesh of Uruk can't stop grieving the loss, and stressing over his own mortality, after his best friend Enkidu dies prematurely. "Must I die too?" he weeps. Siduri the tavern keeper warns Gilgamesh that "the life of man is short. Only the gods can live forever,"[1] but Gilgamesh, undeterred, sets out on an epic journey to seek the secrets of the gods and the key to immortality.

He finds the immortal man Utnapishtim, a Mesopotamian Noah who survived a great flood after being instructed by the god Enki to build a boat and fill it with animals.* After much coaxing, Utnapishtim finally tells Gilgamesh where at the bottom of the sea to find a magical, youth-restoring "wondrous plant." Gilgamesh locates the plant and is taking it home when it's stolen by a sneaky snake. The snake becomes young again, but with no way to get a replacement, the humbled Gilgamesh heads home, finally accepting his own death as inevitable.

Like Gilgamesh before his enlightenment, I have long wondered how it can be that our lives form such a brief arc, shifting so quickly between two crying moments at the hospital. What cruel hoax determined that our muscles start to lose their fiber when we are only in our twenties, that

---

* For those who still think the Bible is an original work…

most of our bodily functions peak in our late twenties, that our chance of death doubles every eight years beginning at age thirty, and that our cells begin losing their ability to repair dangerous mutations starting at around forty? I am not alone.

For as long as humans have been around, we have struggled with our mortality. Even if it forces us to take stock of our aspirations while we have the capacity to achieve them, mortality is at best a mixed blessing.

Lacking the tools to fight back, our ancestors could never really suppress the drive to steal immortality from the domain of the gods. That's why so many cultures have been obsessed with extending life and overcoming death.

In the Old Testament, Methuselah was the one who beat biology. According to Genesis 5:27, Methuselah lived to the ripe age of 969. Because the Bible doesn't say much else about Methuselah, we don't how he did it but are told that his forebears lived between 895 and 962 years, so genetics is a good guess. Regardless, Yahweh seems to have changed his mind when he says a few verses later in Genesis 6:3, "My Spirit will not put up with humans for such a long time, for they are only mortal flesh. In the future, their normal life span will be no more than 120 years."

For the ancient Chinese, who spent centuries trying to generate elixirs of life, the secret was the famous *lingzhi*, the supernatural mushroom of immortality allegedly found high in the mountains. For the Indians, amrita, also known as soma, was a drink made of an elusive, mountainous plant that, according to the Rigveda holy text, let a person live forever.

The advent of the Scientific Revolution in Europe brought a new rationality and new hope to humanity's search for immortality.

In 1896, a Russian-born French doctor, Serge Voronoff, set sail for Egypt after studying for years with the Nobel Prize–winning father of transplant medicine, Dr. Alexis Carrel. Witnessing how the bodies of Egyptian eunuchs seemed to be wasting away, Voronoff concluded that their lack of testicles was denying them the glandular secretions their bodies needed to maintain vitality. Drawing on what he had learned from Carrel about the possibilities of grafting body parts between humans,

Voronoff came up with the creative idea of splicing monkey testicles into human males to supercharge glandular secretions and cure all sorts of diseases, increase vitality, and extend life. "The sex gland," Voronoff wrote in 1920, "pours into the stream of the blood a species of vital fluid which restores the energy of all the cells, and spreads happiness."[2]

By 1923, Voronoff's procedure was in such demand that a special reserve needed to be set up in Africa to contain all of the captured male monkeys awaiting castration to fuel this increasingly popular and massively expensive procedure. After the men with the monkey testicles grafted into their scrotums failed to see much improvement, Voronoff doubled down on his claims. If the monkey donors and the human recipients of the grafts were matched by blood type, he asserted, the humans could live to 140. When none of these claims materialized, the procedure fell from grace.

Monkey testicles were out, but a whole slew of magical procedures and potions—from sewing goat testicles into a person to drinking special "elixirs of long life"—flooded the market at the turn of the twentieth century. All were eventually discarded when they didn't work. Then a strange thing happened. Even though none of the magic potions were doing much of anything, the average human life span started to increase at a faster rate than ever before.

For most of our history, the average human life span has been painfully short. There were lots of ways for our nomadic hunter-gatherer ancestors to die. If you didn't die in childbirth or of the countless infections and diseases in your early years, chances were relatively high that some type of accident or predator or conflict would eventually get you. That's why the average life span for early humans was only about eighteen. In spite of all the advances, the average life span in Roman times was a mere twenty-five. By 1900, it was only forty-seven in the United States, by then one of the more advanced countries in the world.

Average life span doesn't mean that the wise elders of Roman times were twenty-six and everyone in the United States a hundred years ago dropped dead at forty-eight. It means that, if we add up the number of years everyone lives in a given era and divide by the number of people, we arrive at the average. So, if two children are born the same day but one dies immediately and the other lives to eighty, their collective average life span is forty. If infant mortality is high, average life span across the population is low, even if decent numbers of people are living long lives.

But over the course of the twentieth century, advances in health care, sanitation, workplace safety, public health, and nutrition both pushed up the average life span like never before and massively increased the number of old people per capita in the population. Today, the average life span is 71.4 globally, 79 in the United States, and 85 in Japan, even though it is still only around 50 in many of the poorest African countries.

This increase in average life expectancy in the developed world over the past century averages out to about an additional three months per year. The number of people living past 100 has increased by almost two-thirds in the United States and quintupled in the United Kingdom over just the past three decades. Japan, which recorded only 339 centenarians in 1971, today has over 75,000. Pew estimates the global population of people living past one hundred will increase from around 450,000 today to nearly 4 million in 2050.[3]

As the possibility of living longer has become our new reality, our expectations of how many years constitute a full life have changed.

It's a good bet that the families of early humans who had died in their forties on the African savannah never felt they'd been robbed. At that time, living to forty was pretty good in light of all the dangers that could cut short someone's life. Today, losing someone in their forties feels like a crime against potential. But when someone dies today in their nineties, an extremely rare occurrence for our ancestors, most people feel it's about right. When asked in 2013 how long they wanted to live, 69 percent of Americans offered numbers between seventy-nine, the current average American life expectancy, and one hundred, with a median age of ninety.[4] "For everything there is a season," Ecclesiastes says, and the season of life feels to most people like it ought to last around ninety years.

But what if more people were living healthy lives to 120 or 130? Would people whose parents, spouses, and friends died at ninety feel like their loved ones had lived a full life or would they feel as robbed as we do today when someone dies at sixty? Would we benchmark longevity based on how long our grandparents lived or instead expect to live about as long as our neighbors and friends? There is nothing magic about the eighty-year life; it's just what we happen to know now at this particular point along the continuum of our evolutionary trajectory. When that changes, so will our expectations.

Even if our perception of biology is malleable, we don't really know how malleable our actual biology of aging could potentially be. But the new tools of the genetic and biotechnology revolutions are giving us a fighting chance to keep pushing the limits of both our total life spans and our health spans, the vigorous periods of our lives.

As a first step in exploring how long we might eventually be able to live, we need to understand what aging is.

For a process that people understand so well intuitively, aging is remarkably complex. Scientists can't agree on a uniform definition of aging because aging is not one thing. It's probably a combination of many different systems in the body all decaying at different rates. Some scientists have thought of aging as a series of changes that make an organism more likely to die, others as a progressive decline in its ability to do things, others as an increase in inflammation levels or oxidative damage in the body, and still others as a decline in the body's ability to activate the stem cells needed to keep cells in good repair.

Whatever the best definition, aging is the leading cause of death among humans because the diseases that kill us most are all diseases of aging. Heart disease, cancer, and chronic lower respiratory disease account for over half of all deaths in the United States and are three of the top four causes of death globally. These chronic diseases have become a relatively greater threat to us as we've learned to better handle the infectious diseases that plagued so many of our ancestors. Because the chronic diseases correlate with age—the older you are, the more likely you are to get them—curing any one does not help all that much. If one disease of aging doesn't get you, another one soon will. Eliminate all cancer from the United States, and life expectancy only goes up a little more than three years. This leads to the conclusion that if we really want to extend our healthy life spans, we'll need to start worrying relatively less about countering each disease of aging and more about slowing aging itself.

Given the maddening number of parts and systems within the human body, addressing individually the special and unique aging process of each system of our body might prove impossibly difficult.* But if aging is significantly a unified experience with some central mechanisms governing the whole process, conceivably there ought to be a way to slow the aging of an entire organism, including each of its parts. A first step toward assessing if this is the case would be to find ways to measure systemic aging.

Everyone ages differently and at different speeds. We all know people who are chronologically young but still seem or look old. We also know people who are old but seem young. At least superficially, that's the difference between chronological and biological age. While chronological age measures the number of years since our birth, biological age seeks to assess the many genetic and environmental factors that make us age differently.

If I look at your driver's license, I can easily tell you how old you are. By knowing how old you are, I can make a pretty good guess about how healthy you are and how much longer you will live. But I won't be able to tell you how relatively young you are for your age or how long you might live compared to other similarly situated people of the same chronological age without more information about your personal health status.

While chronological age is straightforward and easy to measure, biological age is not. You may seem younger than you are, but we'd need to be able to measure your biological age before and after any kind of antiaging treatment to determine whether this intervention is actually working.

Since the 1980s, researchers have been working to define what this

---

*   I've always wondered why people who believe in the "intelligent design" theory of evolution are not asking why our designer, if he or she is so intelligent, gave us so many fragile individual parts.

type of benchmark for biological aging might look like. More recently, the American Federation for Aging Research, AFAR, established a goal of identifying a biomarker of aging that could accurately predict the rate of aging, measure the general aging process instead of the impacts of disease, be repeatedly testable without harming the person, and work both in laboratory animals and humans.[5] Achieving this is easier said than done. Biomarkers of aging likely include a dizzying and often overlapping list of genetic, metabolic, and other factors that are extremely difficult to measure and, even when measured, are tough to assign specifically to aging.

Recently, however, researchers are starting to make more progress. Studies have suggested that epigenetic markers measured in blood,[6] the length of the genetic "caps" at the end of chromosomes called *telomeres*,[7] walking speed,[8] observable facial aging,[9] and many other factors are preliminary biomarkers of aging that could, in the future, come together to help us solve the riddle of biological aging. The California start-up company BioAge Labs is using AI to make sense of the sequenced DNA from and metabolic analysis of blood cells to identify complex biomarkers of aging in the blood. Blood stored for decades in European blood banks, where the biomarkers and the life and death record of the blood donors can be matched, is proving a uniquely valuable resource.

Being able to measure biological age will help us assess efforts to manipulate it, but we'll still need to look for clues about how long we can live and what we would need to do to live longer and healthier.

※

The good news for thinking we might be able to overcome today's mortality limits is that, within limits, evolution doesn't seem to care much how long we live.

If our ancestors had faced a problem of too many babies being eaten by predators, our offspring might have eventually been selected with exoskeletons like lobsters. If too many parents were being eaten and not able to care for their children, our infants would have eventually become as self-sufficient as loner Komodo dragon babies, who scurry away after hatching to avoid being eaten by their mothers, or the Lambord's chameleons of Madagascar, who hatch each season all alone because the entire adult population dies off after the females lay their eggs.

If grandparents, on the other hand, who were around far less in the early days of human development, got devoured by predators, it would be sad for the family and community but probably not have had a huge evolutionary impact on our species as a whole. Elders are and have always been incredibly useful and important carriers of traditions and critical life lessons—and some smaller proportion of elders were likely needed to look after children while mothers were out foraging for food—but evolution was largely unaffected by whether most lived forty, fifty, or eighty years. The capabilities of babies and parents are essential to human evolution, grandparents less so.[10]

If we consider possibilities for changing a human attribute affirmatively selected over hundreds of millions of years—breathing is a good example—we push hard against the weight of evolution itself. But trying to tweak an attribute that evolution has somewhat ignored should, theoretically at least, give us some preliminary cause for hope.

But while we haven't necessarily been selected for longevity, we certainly have been selected for our ability to survive hard times. Many times over the billions of years of our evolution, our ancestors have teetered on the brink of starvation. Genetic analysis suggests that about 1.2 million years ago our prehuman ancestors dwindled to a mere twenty-six thousand people. Then, about 150 thousand years ago, our ancestor population may have shrunk to as little as 600 people barely hanging on to life in the southern tip of Africa. After the ash spewing from a massive Sumatran volcano 70,000 years ago dimmed the sun for six years, the number of total humans may have again shrunk to as little as a few thousand souls.

With each of these accordion-like swings, only those of our ancestors with the greatest ability to do without nourishment survived. Everyone else withered away. These extreme survivors were then able to pass on their genetic predisposition for surviving tough times through the generations to us. This has left us, like many other species who have faced these same types of challenges, with a built-in heartiness that turns out to be quite useful in our quest for longevity.

Scores of studies have shown repeatedly and for decades that calorie restriction, or CR, extends the life of yeast, flies, worms, mice, rats, and other organisms. Studying whether CR extends the life of longer living mammals like us is more difficult because our longevity makes human studies too darn long and our naturally erratic behavior makes it all but impossible to force human subjects to meticulously avoid eating Ding Dongs for their entire lives. Starting in the 1980s, however, scientists set up a couple of studies seeking to measure the impact of CR on macaque (also known as rhesus) monkeys, which share 93 percent of their DNA with us. The CR monkeys in both studies lived an average of three years longer and remained on average healthier than their peers living in the same conditions but consuming more calories. At least four of the monkeys in one of the studies exceeded all previous records for the longest-lived macaques in captivity.[11] Promising.

A U.S. National Institutes of Health–funded study called the *Comprehensive Assessment of Long-Term Effects of Reducing Intake of Energy*, or CALERIE, convinced thirty-four people to reduce their overall calorie intake by 15 percent for two years. These people agreed to submit to weekly blood, bone, urine, internal body temperature, and other tests, and even spend twenty-four-hour periods in sealed metabolic chambers so that their breath could be measured for its ratio of oxygen to carbon dioxide. Based on all of this input, the researchers concluded that the 15 percent decrease in calorie intake translated into 10 percent lower metabolism. Although the study could not determine whether the still-living subjects would live longer than they would otherwise have, they concluded that lower metabolism led to a "decreased rate of living," or less wear and tear on the cells, and the possibility of longer and healthier lives for the human equivalent to the macaque monkeys.[12]

Another place to look to understand how long we might have the potential to live is at the oldest Homo sapiens around us. Gilgamesh and Methuselah notwithstanding, the oldest person reliably recorded was a remarkable French woman named Jean Calment.

Born in the French town of Arles in 1875, young Jean encountered Vincent Van Gogh when she was twelve and married her cousin in 1896, who died in 1942 when Jean was sixty-seven. She was just getting started.

When Jean was ninety, a young lawyer named Francois Raffray convinced her to sell her apartment to him on a contingency contract. Jean would still live in the apartment, and Raffray, according to the contract, would pay her 2,500 francs a month, the equivalent of $500 dollars today, until she died, when he would take possession of the home. Little did he know...

Eating two pounds of chocolate a day, Jean rode her bike around the city until she was 100, remained extremely active past 110, and only began to slow down after a fall at 115. By then, she had become a local celebrity whose fame was growing with the passing years.

Raffray, meanwhile, was contractually bound to keep sending the monthly checks, whose value now far exceeded that of the apartment. After he died at age seventy-seven, his family was still legally required to keep paying for an additional year until Jean Calment died in 1997 at the age of 122, two years beyond Yahweh's declared upper limit from Genesis 6:3 (but then again Yahweh doesn't appear in the Bible to have known about chocolate).

Books have been written about why Jean lived so long. Famously unflappable, she attributed her longevity to her low levels of stress and positive attitude, but genetics also played a key role. "I've only had one wrinkle in my life," Jean once famously said, "and I'm sitting on it."

Jean Calment's colorful story tells us what might be the upper limit of possibility, based on today's biology and intervention options. To be more systematic, we can also try to understand large groups of super-agers like Jean to see what allows them to live that long. That's exactly what Nir Barzilai is doing.

Director of the Institute for Aging Research at the Albert Einstein

College of Medicine and one of the world's great experts on aging, Nir has been recruiting and studying a large and growing number of Ashkenazi Jewish centenarians* in the greater New York area for years. There are many words that could be used to describe Nir—spritely, cherubic, infectiously positive—but none fully captures his essence. Although described by some as an elfin Austin Powers lookalike, Nir is a giant in the field of aging.

He's encountered some truly remarkable people along the way, like Irving Kahn, a passionate learner who worked as an investment banker until his death at 109. Irving's siblings, collectively the longest living siblings in recorded history, lived to be 109, 103, and 101. Irma Daniel, another of Nir's centenarians, was a sharp-witted survivor of Nazi Europe who held court in Hoboken, New Jersey, until the age of 106. For each of these people, Nir and his colleagues are taking full life and health histories, sequencing their genomes, and conducting a battery of tests to determine what might be the "special sauce" for living so long. Although the research is ongoing, it has already made clear that while healthy habits can help people live somewhat longer lives, the best way to today live past one hundred is to have the right genetics.

These supercentenarians don't just live longer, they live healthier longer. Most tend to have most of their faculties late into their lives and only die after relatively quick and compressed periods of illness.[13] A Scripps Research Institute study of the sequenced genomes of over fourteen hundred "wellderly" people over the age of eighty similarly found that genetic variants in these people seemed to help maintain their cognitive health and protect them from major chronic diseases.[14] Although popular perceptions of people living into their one hundreds imagine them hooked up to ventilators, the cost of health care for the average person dying past one hundred is only 30 percent that of the average person who dies in his or her seventies.

Nir and other researchers found that while many of the centenarians

---

* Centenarians are people who live to be age one hundred and beyond.

would be expected, based on their age, to pick up diseases like Parkinson's, Alzheimer's, and cardiovascular disease at higher percentages, they somehow didn't.[15] His team zeroed in on the gene ADIPOQ, common in most people but absent in many of the super-agers, which seems to protect them against inflammation of the arteries. Other researchers have identified tens of genes that, in one expression or another, seem to protect against brain disorders[16] and inflated cholesterol levels,[17] provide additional protection against Alzheimer's,[18] and increase life span more generally.[19] More genes associated with longevity are being identified at a rapid clip.

They say if you'd like to live to ninety, eat well, relax, sleep, and exercise. If you want to live past one hundred, choose your parents wisely. But this maxim could very well become OBS—overcome by science.

Identifying more of the genes that increase a person's potential to live longer and healthier will allow us to introduce some of these genes into people via gene therapy or, more likely, figure out what the genes are doing and find a way to mimic that.

We all know that no matter what our genetic predisposition, we can live longer and healthier lives if we make smart lifestyle choices. Although lifestyle choices often feel like a separate category from genetics, they are not. Our lifestyle choices significantly impact the epigenetic instructions orchestrating how our genes work. Understanding which lifestyles facilitate the most optimal expressions of our genes, therefore, tells us not only what we might do to live healthier lives but also, at least for the cheaters among us, how we might trick our genes and our biology more generally to give us credit for the smart choices we have not made.

Trying to crack the code of what types of backgrounds and lifestyles deliver the greatest benefit, National Geographic Fellow Dan Buettner poured over global life-expectancy records, looking for the world's statistically longest-lived populations. He labeled the places he found—in Icaria (Greece), Loma Linda (California, United States), Nicoya (Costa Rica), Okinawa (Japan), and Sardinia (Italy)—blue zones and set out to uncover their common denominators.

In a 2005 *National Geographic* cover story and 2008 book titled *The Blue Zone: Lessons for Living Longer from the People Who've Lived the Longest*, Dan described how people in blue zones do the following nine things:

1. They have moderate, regular physical activity woven into their lives that nudge them into moving every twenty minutes or so—not necessarily going to the gym but getting around their towns or villages to take care of necessities;
2. They can articulate their life purpose, or *raison d'être*—what the Japanese call *ikigai*;
3. They honor sacred, daily rituals that help them live relatively low-stress lives;
4. They only consume a moderate amount of calories per day;
5. They have plant-based, but not necessarily vegetarian, diets, mostly consisting of whole grains, tubers, nuts, greens, and beans;
6. They consume alcohol only moderately, if at all;
7. They have engaged spiritual or religious lives;
8. They have active and integrated family lives; and
9. They were born into or joined committed circles of devoted friends and are embedded in regular, active, and highly supportive social communities.[20]

People in blue zones weren't necessarily becoming centenarians at higher rates than everyone else but they were, on average, living longer and healthier longer. So, living like people in blue zones can help us live longer than some of our neighbors but not longer than Jean Calment (who ate more chocolate than tubers and nuts!). In addition to looking at how and why certain people and communities live longer than others, another way of cracking the genetic code of mortality is to look for clues in the animal world by examining related animal species with significantly different life spans.

The average mouse, for example, can live up to about three years in the wild and four years in captivity, but the remarkable naked mole rat, a close relative, can live up to thirty-one. Less mature writers than I have described the naked mole rat as looking like a penis with two sharp teeth but I, a paragon of restraint, will just let you decide for yourself.

Photo courtesy of University of Rochester.

Native to Ethiopia, Kenya, and Somalia, these underground-dwelling, highly social, nearly hairless mammals live far longer than would be estimated based on their size. On average, larger species tend to live longer than smaller ones even though smaller versions of the same species tend to live longer than larger versions.[21] Naked mole rats live more of their lives in good health than most other comparable creatures, repair any genetic damage more steadily, apparently feel little pain, and are immune to cancer. For these reasons, scientists are increasingly deploying genome sequencing, big-data analytics, and other advanced tools to try to understand the secret of the naked mole rats' success.

Calico, for example, Google's San Francisco–based life-extension company, maintains one of the world's largest captive colonies of naked mole rats to see if it can uncover biomarkers of aging and unlock the secrets of naked mole rat longevity.[22] Other studies are already beginning to generate hypotheses for what's helping these creatures live so long and healthily, with an eye toward ultimately expanding human longevity, cancer resistance, and health span.

One study hypothesized that naked mole rat genetics make their cells produce high levels of a protein called *HSP25* that serves almost as an automatic spell-check function, eliminating other faulty proteins in cells before they can cause a problem.[23] Another found that naked mole rats have four pieces of ribosomal RNA—the tiny structures that translate DNA code into instructions for the cell to manufacture proteins—instead of the usual three for most other multicellular life forms like us. For a reason that scientists don't fully understand, the four-part naked mole rat RNA structures makes significantly fewer translation errors than the three-part structures in other mammals.[24]

Researcher Vera Gorbunova explains one logical theory for why naked mole rats live longer and healthier than we might expect, compared to the biology of mice: living underground protects the naked mole rats from predators. Because their likelihood of being killed or eaten is high, mice invest evolutionary energy into fast reproduction. Naked mole rats, on the other hand, protect themselves from predators by living underground so can invest more evolutionary energy into living longer lives.[25] Gorbunova and her scientist husband Andrei Seluanov are now exploring whether they can engineer some of the unique genetic structures of the naked mole rats into mice to see if those mice live longer.

Other examples of related animals with dramatically different life spans are the hard clam, *Mercenaria mercenaria* in scientific lingo, and its closely related Icelandic quahog clam, or *Arctica islandicaI*. The hard clam can live about forty years. Its cousin the Icelandic quahog clam can live over five hundred. No one really knows just how long a quahog clam can live because the Icelandic researchers who in 2006 discovered a 507-year-old quahog (they calculated how old it was by counting the growth rings on its shell) ended up killing it accidentally when they brought it up from the ocean floor and opened it to measure its age. Oops.

A major study used a battery of tests to compare these quahog clams to their shorter-lived hard clam relatives and found that the Icelandic quahogs' remarkable resistance to oxidative stress—the damage done to cells when they are exposed to electronically unstable molecules called *free radicals* that accrue in all animals over time—was likely a major factor in their extreme longevity.[26] This finding pointed researchers toward exploring oxidative stress as a potential systemic element in the aging process of other animals, including us.

My favorite example of an animal that shows us how little we know about aging and what might be possible is *Turritopsis dohrnii*, the immortal jellyfish.

It's hard not to get excited about the *Turritopsis dohrnii*. Like all jellyfish, it starts out as a fertilized egg, out of which a larva emerges that then sinks to the ocean floor. There, it grows into a colony of polyps that spawn multiple, genetically identical jellyfish. But when faced with physical adversity or a lack of food, the *Turritopsis dohrnii,* unlike other jellyfish, turns from being an adult jellyfish back into a polyp, kind of like an adult human turning back into an embryo.

Comparing the biology of the immortal jellyfish to the biology of other subspecies of non-immortal jellyfish has helped scientists explore how the next generation of research into how induced pluripotent stem cells might help unlock secrets of cellular regeneration and rejuvenation, keys to extending human life.[27]

Humans and jellyfish separated in our evolutionary journey about six hundred million years ago. That's a lot of years. But understanding more about what enables naked mole rats to live long and cancer-free, for quahog clams to slow their metabolism and protect their cells from oxidation, and for immortal jellyfish to continually regenerate themselves is already generating clues about how aging happens and how the process might be manipulated for us.[28]

X

Most of my childless friends often come away, like me, from visits with other people's small children with the same wind-blown look on our faces. These awesome kids with so much darn energy, our internal dialogue goes, are great for an hour or two but would be overwhelming were they our full-time responsibility. (This feeling, of course, largely explains why some of us remain childless!) But the biological underpinnings of all this energy has big implications for aging.

All animals live on energy, but our ability to process the energy resources we have are limited. At every moment, our cells need to figure out how to allocate the energy they have. There are two main options: growth or repair. When we are young, we need lots of energy to grow into ourselves. That's why young people, for example, have such a high metabolism, seem hyperactive, and burn a lot of calories. But burning this much energy also comes at a cost. Like constantly driving a car at full throttle, it wears out our cells. But cells have another option: to allocate energy more conservatively toward repair, kind of like the quahog clams, or like we do in the face of calorie restriction or other stress. This is equivalent to when your computer shifts to screensaver mode to conserve electricity or when your grandmother drives her car twenty miles an hour and carefully parks it in her garage.

Cells do this by regulating the uptake of glucose, the sugar we need to survive. Normally, cells use glucose aggressively for growth, but in times of scarcity, our cells shift into repair mode. The animal models showed that when the roundworms, fruit flies, and mice shifted from growth mode to repair mode, they were not only better able to survive hardship but also lived longer and healthier lives. When scientists like biologist Cynthia Kenyon (then a scientist at University of California, San Francisco, but now with Google Calico) used new tools to turn on and off the genetic switches that these experiments identified, they found that a few genetic changes to the Daf-2 and Daf-16 genes could shift cells from growth to repair mode and double the life spans of C. elegans roundworms.[29] This led to the increasing realization that the aging process is regulated by the equivalent of knobs that can be turned.

Although the idea that we have some kind of Wizard of Oz inside us turning knobs to determine how long we live seems strange to many people, it really shouldn't feel so strange. Somewhere in our biology we have the ability to turn back time on a cellular level, kind of like the immortal jellyfish, evidenced every time we reset the genetic clock by making a baby.

But even if we better understand the underlying mechanisms of aging and believe that reprogramming our biology to extend our healthy life spans is possible, the question of how we to do it remains.

When reading a book about science, the last thing most people want is yet another reminder to get off their duffs. But the hard science of longevity makes abundantly clear that smart lifestyle choices are the best first way to hack our own epigenetic signals.*

Repeated studies have shown that these types of behavior changes also can lengthen our telomeres, the stretches of DNA at the end of our chromosomes. The telomeres serve to protect the integrity of the genetic data inside the chromosome, like plastic tips protect shoelaces. Our telomeres get slightly shorter each time our cells divide, becoming less and less able to protect our genetic data from harmful mutations. Although it's not entirely clear whether shortened telomeres are a cause or an effect of aging, shorter telomeres are associated with faster aging and higher risks of age-related disease.[30]

But even if we don't live like the people in blue zones, we still need to recognize that our biology and our lifestyle choices are not different things in the context of our health and longevity. Instead, they are the interconnected points along the spectrum of ourselves. Our lifestyle choices help realize the potential of our biology. Our biology helps determine the

---

*   This is not a self-help book, but if you want to live longer and healthier, I recommend that you incorporate as many of the lessons of Dan Buettner's blue zones as you can: exercise or at least get moving every day, find your spiritual center and special purpose in life, decrease your stress level, switch to a mostly plant-based and not-too-high-calorie diet, invest in your social and family life, and stop drinking too much alcohol. I would also be remiss, or at least in trouble, if I did not mention that my brother Jordan has written a fantastic book on the systemic health benefits of exercise: Jordan Metzl, *The Exercise Cure: A Doctor's All-Natural, No-Pill Prescription for Better Health and Longer Life* (Rodale, 2013).

extent to which these types of choices can help us. And because we are humans, most of us are looking for new ways to beat the system.

We know that our bodies have a built-in genetic survival mechanism that has protected our ancestors in past times of scarcity and stress that can potentially be exploited for our advantage. Because much of the preliminary research on the genetics of extending life and health span seeks to understand how the human body responds to calorie restriction and scarcity, it's logical to ask if we shouldn't all just start living on a calorie-restricted diet. It could conceivably help, but I don't necessarily recommend it. Even though the animal studies are pretty convincing, it's yet not fully proven that calorie restriction extends human life and health span, possibly because testing this hypothesis on humans is so difficult.

Nevertheless, calorie-restriction societies have popped up around the United States and world over the past couple of decades, bringing together people committed to significantly cutting their calorie intake in the hope of extending their lives. For most everyone else, however, cutting calories to about 1,500 calories a day for men and 1,800 for woman (compared to the USDA recommended two thousand calories for active adult women and twenty-five hundred for active adult men) over the course of an entire life is hardly an appealing prospect and might not even be as healthy overall as a balanced diet.[†]

But a recent study found that significantly reducing calories just five continuous days a month, for two months a year, might elicit the same conserving resources response from the body and give people the same health benefit as cutting calories every day.[31] The case, based on animal studies for intermittent fasting, is not as compelling as continuous calorie restriction, but fasting once in a while or for blocks of time each day might seem a more palatable way for most people to get at least some of the potential benefits of calorie restriction without feeling miserable all the time.

Perhaps a more appealing option, at least for some people, is exercise.

---

†   As someone addicted to my morning cup of hot chocolate and deeply passionate about salted caramel soufflé, even I can't in good faith sign on to such a restrictive regimen.

I'm an Ironman triathlete and ultramarathoner and one of those crazy people who think doing these types of activities is actually fun.* For most people, however, exercise is a chore. If offered a pill that elicited the same protective genetic and metabolic response in their bodies as if they had exercised, most people would not only just say yes, they would pay huge sums of money or raid their local pharmacy to get it. But by triggering the same cellular shift to repair-and-conservation mode, exercise does all of those things for free.

A recent comprehensive study followed more than 650,000 people for an average of 10 years and analyzed nearly 100,000 death records in an attempt to quantify the input versus output benefits of exercise. The study found that 75 minutes a week of moderate exercise, like brisk walking, led to almost 2 years of added life expectancy compared to sitting on your couch. The benefits went up from there. Adding 2.5 to 5 hours of exercise a week added 3.5 years. An hour a day, or 450 minutes a week, added 4.5 years to life expectancy.[32]

Even with these odds, not everyone is willing to restrict calories or exercise religiously, and even those who do both will still likely want whatever additional benefits might help them live even longer and healthier lives. The good news for all of us is that drugs mimicking the function of the genes we wished we'd inherited or the impact of the calorie restriction or exercise we wished we'd done will soon likely be available thanks in part to the new tools of the genetic revolution.

X

One reason that, unlike the naked mole rats, our ability to repair damaged DNA declines as we get older is that the levels in cells of a molecule called *nicotinamide adenine dinucleotide*, or NAD+, decreases as animals like us

---

* Alter completing a nineteen-hour ultramarathon in the rain forest of Taiwan, someone back at my Taipei hotel told me I must be getting paid a lot of money to participate in the race. He was incredulous when I told him I actually paid an entrance fee.

age. The NAD+ molecules amplify the activity of a group of seven special genes called *sirtuins* to augment their ability to repair damaged DNA. The more NAD+ in the cells, the better able they are to fix these problems.

An obvious approach to potentially fixing this problem might be inserting NAD+ molecules into cells, but the NAD+ molecule is too large to make it through the cell's outer membrane. Instead, scientists figured out a way to use smaller, precursor molecules called *nicotinamide mononucleotide*, or NMN, and *nicotinamide riboside*, or NR, that are small enough to make it through. Once in, the NMN and NR bind with a molecule already inside the cell and make NAD+.

When Harvard scientist David Sinclair and his colleagues genetically engineered older mice to express higher sirtuin levels, or used NMN to increase their NAD+ levels, they found the mice had better organ function, endurance, disease resistance, blood flow, and longevity than other mice the same age.[33] In many ways, boosting the NAD+ levels tricked the mouse's cells into shifting from growth to repair mode. Although human trials are only just beginning, the market for NMN and NR pills, sold in the United States as unregulated supplements, is exploding.

Another drug that shifts the balance of cellular activity from growth to repair mode is the seeming wonder drug metformin. Doctors have prescribed metformin since the 1950s, but the history of its essential ingredient actually goes back much further. In its earlier incarnation, medieval herbalists used a plant known as French lilac, or goat's rue, to treat a multitude of ailments, including frequent urination, today recognized as a telltale sign of diabetes. In 1994, the U.S. Food and Drug Administration approved metformin as a treatment helping diabetics keep their blood sugar levels in check. Because it's been around so long, metformin is known to be relatively safe, is not patent protected, and costs about five cents a pill. Today, Americans fill about 80 million metformin prescriptions every year.

With so many people taking the drug, doctors around the world began first observing and then studying some of the surprisingly positive effects metformin was having beyond controlling diabetes. A 2005 study found that metformin reduced the risk of cancer among diabetics.[34] In a 2014 study comparing metformin to another diabetes drug, researchers found that the metformin-taking diabetics didn't just live longer than the other diabetics taking the other drug but also lived longer than the control patients who didn't have diabetes at all.[35] The same year, a Singaporean study concluded that metformin halved the risk of cognitive impairment among older diabetics.[36] A slew of mouse studies also showed the same types of miraculous results. Male mice given the drug lived an average of 6 percent longer than those who weren't, and all mice taking metformin had fewer cancers and less chronic inflammation than the controls.[37]

The combined force of these findings pushed scientists toward the inevitable hypothesis that metformin was not just impacting a few individual diseases but instead having a systemic effect on entire organisms. It was, in effect, making regular people more like the centenarians with their genetic predispositions for longer and healthier lives or the people living in blue zones. This makes logical sense. Insulin tells our cells it's time to grow. When we have too much of it, either by gorging at the dessert bar or being a diabetic, our cells overinvest in growth at the expense of repair. When we moderate our insulin uptake through diet, exercise, calorie restriction, or metformin, our cells shift back into repair mode, like Cynthia Kenyon's Daf-mutated roundworms. This helps us reduce oxidative stress like the quahog clams and better fight disease and live longer and healthier like the naked mole rats.

To answer the question of whether metformin might be a systematic drug that helps enhance healthy life spans among nondiabetic populations, Nir Barzilai and his collaborators are now exploring how metformin might delay the onset of multiple age-related diseases and stem decline in older people's physical performance, cognitive clarity, and quality of life. Proving for the first time that a single drug like metformin can target multiple diseases of aging simultaneously would revolutionize the field of aging.

NAD+ boosters and metformin may be among the first drugs that target aging, but they will certainly not be the last. Another drug that's proven to extend the life of all animals tested in studies around the world is the miraculous drug rapamycin.

In 1965, Wyeth Pharmaceutical scientists visited tiny Easter Island in the Pacific looking for soil bacteria that might have antifungal properties. Among the thousands of samples they collected, one contained unique bacteria that secreted a compound allowing the bacteria to absorb as many soil nutrients as possible while stopping other competitive funguses from growing. The scientists named the compound *rapamycin* in recognition of the island's local name, Rapa Nui.

Rapamycin's natural ability to slow the proliferation and growth of targeted cells made it an ideal immunosuppressant, perfect for preventing people's immune systems from rejecting transplanted organs. But then doctors started noticing that animals and some human transplant patients taking rapamycin seemed to be healthier than similarly situated animals and people taking other drugs. They soon figured out that rapamycin regulated the metabolism of cells and triggered the same kind of shift-to-repair-mode signaling as happens when calories are scarce and which we also saw with metformin. They named the protein targeted in this process *mTOR,* or *mammalian target of rapamycin.*

With almost every study, the list of rapamycin's almost magical capacities increased. By regulating cell growth, rapamycin proved extremely useful in treating specific diseases where out-of-control metabolism and cell-growth was the problem, including cancer, diabetes, heart and kidney diseases, neurological and genetic disorders, and obesity.[38]

The quahog clam can live over five hundred years because it slows its metabolism and shifts the energy of its cells from growth toward repair mode, so it makes sense that by essentially triggering the same shift a drug like rapamycin could impact longevity. In studies run on species after species—yeast, flies, worms, mice, and rats, to name but a few—ingesting rapamycin led to about 25 percent longer lives across the board, an astounding feat.[39] Leading longevity and rapamycin expert Matt Kaeberlein is now launching a major multiyear effort to study the life and health-span impact of rapamycin on companion dogs as a step toward better understanding how the drug might work on humans.[40]

In spite of all this promise, there's a good reason why scientists weren't immediately hailing rapamycin as the human fountain of youth. Having our immune systems suppressed when we are getting organ transplants is making the best of a bad situation. We benefit from the transplanted organ but pay the price by turning off our immune system, making us more vulnerable to dangerous viruses and bacteria we would otherwise be able to fight off. Rapamycin also can have other dangerous side effects for immune-suppressed people, including anemia, hyperglycemia, cataracts, and testicular degeneration.[41] It's still not clear whether healthier people taking rapamycin would experience the same risks.

Researchers are now working hard to try to find way of delivering rapamycin or a derivative that can maximize the compound's remarkable benefits while minimizing its downsides.[42] Pharmaceutical behemoths like Novartis and start-ups like Boston's resTORbio and PureTechHealth are racing to develop drugs using rapamycin to help shift our cells toward repair mode and potentially enhance our health and longevity.

Before human trials are completed, it's not a good idea for people to start taking NMN, NR, metformin, or rapamycin on their own in an effort to extend their life or health spans. Having said that, I've been surprised by how many of the NMN and NR researchers are themselves taking NMN and NR, how many metformin researchers are taking metformin, and even how many rapamycin researchers secretly confess that they are experimenting on themselves by taking rapamycin. And I do think it's likely that many people around the world will within a decade be taking some type of antiaging drug using some or all of these ingredients or derivatives of them, as well as other compounds yet to be identified.[43] Ideally, this pill will be personalized and different for different people, based on gender, age, genetic profile, metabolic status, microbiome diversity, and other factors. But this won't be the only option.

X

When we are young and relatively healthy our cells divide regularly to replace themselves, but as we get older, or face other stresses, some of our cells stop dividing. Rather than just dying and getting flushed out of our systems, these zombie-like "senescent" cells secrete molecules, causing increasing levels of inflammation and tissue damage. The older we get, the more of these senescent cells we accrue. That's not entirely a bad thing, however, because senescent cells also help repress the growth of cancer cells and tumors, which are an increasing danger as we age.

We probably wouldn't want to get rid of all of our senescent cells and lose their benefits, but pruning the level of senescent cells in our bodies, it now appears, might help us function a bit more like when were younger and healthier. After scientists genetically engineered and then activated a transgenic "suicide gene" to pare senescent cells in mice, the mice not only lived 25 percent longer but also regrew lost hair, had stronger muscles and better functioning organs, and experienced increased insulin sensitivity and lower rates heart disease and osteoporosis.[44] Not bad. In August 2018, a team of researchers at England's University of Exeter used three different novel compounds to prune senescent cells in the mitochondria of mice, which then showed significant signs of cellular rejuvenation.[45]

With this research as a backdrop, the race is now on to develop and test a new class of drugs called *senolytics*, designed to prune senescent cells to treat specific diseases of aging and potentially expand both health span and life span. Few humans will sign up, at least for now, to have "suicide genes" genetically engineered into their genomes, but many of us would take a pill that had the same effect, if it were available. With this in mind, Dutch scientists have designed a chain of amino acids that bump up our body's ability to clear out senescent cells. Interestingly, these amino acids influence the same Daf-16 genes Cynthia Kenyon identified in her roundworm studies. Older mice given the compound regrew the luxurious fur of their youth and were able to run twice as far as they could just weeks before.[46]

A related potential approach to countering aging involves manipulating the way our cells recycle their own biomass to extract energy and remove harmful proteins, a process scientists call autophagy. Autophagy serves younger organisms well but can malfunction and cause age-related diseases later in life. A new class of potential treatments for some age-related diseases and for aging itself, therefore, involves the targeting of specific steps of the autophagy process to help older cells start behaving as if they are younger.

Human clinical trials are now just starting to explore the potential impact of senolytic and autophagic drugs on specific diseases correlated with aging as a first step toward treating aging itself. Still, many hurdles would need to be to overcome before these drugs can be safe for human use. We need to make sure, for example, they aren't incapacitating our body's natural defenses against cancer or triggering other unintended negative effects as we potentially mess with billions of years of biological checks and balances.

If you are already amazed by how many potential ways there may be to hack aging, buckle up. We are just getting started.

※

Baby-making is so commonplace that we can often miss the clue about human aging right in front of our eyes. Every time a child is conceived something in our cells is already resetting the biological clock.

Each of our cells, as we know, contains the genetic blueprint of our entire genome. That's why it would be possible to clone a person from a single one of their cells. But we would be in big trouble if each of our thirty trillion cells kept aspiring to be its own person. The reasons this doesn't happen is that, as we've seen, the epigenetic signals in our cells control which of our genes are turned on and which are turned off. As we develop from the single cell of a fertilized egg into the complex beings we become, these epigenetic marks choreograph our cells to become specialized, adult cells like heart cells, skin cells, and so on, each performing their specific functions.

Shinya Yamanaka's miraculous 2006 innovation was that the four *Yamanaka factor* genes could reprogram these specialized cells back into stem cells. The epigenetic marks from the adult cells were being removed, similar to what happens when a child is conceived.

You probably remember from childhood that when a starfish loses one of its arms, a spider loses one of its legs, or a lizard loses its tail, those appendages can often grow back. You probably also remember from every U.S. Civil War movie you've seen that humans don't have the same opportunity. Being able to look deep into the cells of the regenerating animals, however, has enabled scientists to figure out that their cells are reverting halfway between their specialized and stem versions to recreate the arms, legs, or tails. Following the injury, their relevant cells revert back enough to regrow but not so far back to forget what they need to grow into.

With this in mind, scientists started to wonder what would happen if they used Yamanaka factors to turn older, specialized cells into younger versions of the same types of cells, rather than all the way back to stem cells. The genes could conceivably stay the same, but the epigenetic instructions telling the genes how to function would change. Remarkably, partially reprogramming adult skin cells in a dish to become younger skin cells worked.[47] And if individual cells could be made younger, some hypothesized, wouldn't it be possible to use the same type of approach to make entire organisms biologically younger?

The first time that scientists tried this, all of the mice died awful deaths as their adult tissues lost their specialized identities and cancers proliferated. But Salk Institute researchers wondered if the idea was right and just the dosages were wrong. After experimenting for five years trying to find an effective but nonlethal dose, they figured out a way to genetically engineer mice with extra copies of the Yamanaka factor genes that could only be expressed when the mice received a specific activator drug in their drinking water.

After multiple efforts with this moderated, partial reprogramming strategy in mice with progeria, a disease causing rapid cellular deterioration and premature aging, the scientists hit pay dirt. The mice treated with this partial epigenetic reprogramming looked better, had better tissue function, and lived 30 percent longer than other mice with the same disease. "Our study shows that aging may not have to proceed in one single direction," Salk Institute researcher Juan Carlos Izpisua Belmonte said at the time. "It has plasticity and, with careful modulation, aging might be reversed."[48]

There are many dangers and pitfalls ahead when considering cellular reprogramming to slow and ultimately reverse human aging, but this research points the way toward epigenetic reprogramming as a way of turning back the clock for all animals, potentially even us. As with the senolytics, finding drugs that do this will be more feasible than genetically engineered hacks, but this work too has already begun. Big pharma companies like GlaxoSmithKline, Eli Lilly, and Novartis are jumping into the epigenetics treatment field.

But wait, as the 1970s Ginsu knife commercial goes, there's still more.

X

In the seventeenth century, eminent British scientist Robert Boyle hypothesized that "replacing the blood of the old with the blood of the young" might help make significant life extension possible. It seemed a crazy idea to many people, but the first real efforts to test this hypothesis came in mid-nineteenth-century France. *Parabiosis*, a term coined by researchers in the early 1900s to describe the burgeoning efforts to decipher animal biology systems by cutting open two different animals of the same species and sewing them together, comes from the combination of the Greek words *para*, meaning *next to*, and *bios*, meaning *life*. In the 1950s, Cornell animal husbandry professor and gerontologist Clive McCay sewed an older and a younger rat together and found that the older rat's bones became stronger, hinting that Boyle could have been right. Parabiosis remained, however, a rather cruel way of doing research, even among scientists comfortable with the inherent but often necessary cruelty of animal studies.

The modern age of parabiosis began in 2015, when a husband and wife team of Stanford researchers published a high-profile paper describing how, when parabiotically paired with younger mice, the older mice healed faster and had better liver function than other older mice not Siamese-twinned to anybody.[49] Since then, a slew of studies have shown that when old and young mice are joined, the older ones get functionally younger in many ways and the younger ones get functionally older. The paired older mice's hearts get healthier, new neurons are formed in their brains, their memory improves, their spinal cord injuries heal faster, their muscles become stronger, their hair thickens,[50] and they move back into their parents' house and put rock-and-roll posters back up on the walls. Just kidding about the posters, but multiple parabiosis studies suggest that the paired young mice in many ways become old, and the old ones in many ways become young.

There are many theories about why this happens, most involving stem cells and other elements of the younger rodents' blood plasma rejuvenating the older partners. To prove this point and demonstrate that parabiosis wasn't just about mice and rats, Stanford's Tony Wyss-Coray injected human umbilical cord blood into older mice and showed that something in the human cord blood was improving the mice's success at memory tests.

The race is now on to figure what factors in the blood are catalyzing changes like these and how these factors could be isolated as a treatment for aging and its associated diseases. Wyss-Coray's company, Alkahest, for example, is testing whether older Alzheimer's patients will see their condition improve after being infused with matched plasma taken from young donors. Once the rejuvenating elements within blood plasma are more successfully identified, isolated, and cultured, we may well see in the near future antiaging treatments designed to enhance these capacities.

You won't see wealthy older people with sorry younger ones stitched to their sides any time soon, but popular culture is already catching on to this trend. In the popular HBO comedy *Silicon Valley*, a wealthy tech CEO pays a young man he calls his "blood boy" to transfuse his blood straight into the older man. In the not-too-distant future, however, we might well see more people transfusing blood plasma from others or even storing their own blood for self-transfusions later in life.*

---

* Today, red blood cells can be stored for only ten years. But more easily stored cells like skin cells could probably be frozen indefinitely then thawed in the future and induced to become blood cells that could be used in self-transfusions. I explore this possibility in my novel *Eternal Sonata*.

In the medium term—as super-ager studies like Nir Barzilai's continue to identify specific genes and genetic patterns that make a given person statistically more likely to live longer and healthier—parents having children using IVF and embryo selection will be able to implant embryos with the best genetic chance of living healthier longer. Not long after that, parents will have the option to genetically alter their preimplanted embryos to further increase these chances. There may be evolutionary trade-offs in making this decisions that would need to be weighed, but people, as we've seen, will want their children to be optimized for a long and healthy life.

As we live longer, the chance that some of our many parts will break down from overuse will become greater, even if our stem cells can remain as active as we age as they were when we were young. But we will have a widening set of tools to fight back over the coming decades—nanobots scouting through our bodies looking for things to fix, biological 3-D printers building replacement parts to implant, and other genetic tools to turn our own biology into machines upgrading our software and hardware from the inside out.

All of this work to expand the human health span will be sped up significantly if we as a global community invest more, and more smartly, in understanding aging and countering its more pernicious effects. In 2017, for example, the U.S. National Institutes of Health only spent only $183.1 million of its overall $32 billion budget, one-third of one percent, on aging biology, far less than the many billions it spent on cancer, arthritis, diabetes, and hypertension.[51] Even within this small base, the National Institute on Aging spends over half of its budget on Alzheimer's disease. Cancer, arthritis, diabetes, hypertension, and Alzheimer's are all terrible conditions that must be addressed, but eliminating any one of them entirely won't extend most of our health spans by all that much because they are all correlated with age. The older we are, the more likely we are to get all of them. That's why the return on investment for the overall population would be significantly greater if we invest relatively more than we are today in understanding and treating aging itself, rather than in each of its many cruel manifestations.

University of Illinois at Chicago researcher Jay Olshansky and his collaborators tried to calculate the cost savings to society of pushing back the average age at which we are afflicted with the multiple diseases of aging. Based on their 2013 study, intervening to slow the aging process and push back the onset of the multiple diseases of aging by 2.2 years across the U.S. population would result in a whopping $7.1 trillion in savings over a fifty-year period.[52] Put another way, if extending health span across the population proves as possible as it now appears, the anticipated savings could not only pay for a Manhattan Project that targets aging but also cover repairing all of America's decaying infrastructure, provide universal preschool for every American child, and provide clean water to virtually everyone on earth.*

Massive life extension will not be easy, and there will be obstacles before us we cannot yet see. Because the oldest person to ever live was 122, for example, we don't know if there is some type of deadly disease we've never heard of that humans will get at age 123. But with biology becoming ever more malleable, the prospect of at least continuing the rapid expansions in health and life span we've seen over the past century into the next seems very real. Making the same forty-year leap from a global average life span of about thirty in 1900 to around seventy in 2000 will be tough because growing global prosperity has already so significantly reduced the infant and child mortality rates that suppressed the 1900 numbers, but this kind of continued growth is not impossible.

The first step would be to help lower infant mortality and increase overall health and longevity in the most vulnerable parts of the world, particularly in sub-Saharan Africa and South Asia. Even if that is achieved, most of us will still want longer and healthier lives.

Over the past century, average life expectancy in the United States has increased by about three months per year. With all the budding technologies I've described, we might be able to increase that rate to four months,

---

* To address the time-lag issue of spending tomorrow's savings today and add an additional incentive for progress, big countries like the United States could offer health span bonds.

then five, then six months per year. If life expectancy could be made to consistently increase faster than people age, then we would reach what Ray Kurzweil calls "life expectancy escape velocity" and basically be immortal. This is great in theory, but I seriously question we'll ever get there in our current time-limited bodies. This book is about hacking our genetics, which will certainly be possible. Even so, I'm doubtful we have enough wiggle room in our biology to make biological immortality possible.

But maybe our biology could become less of a limiting factor if we merge in new ways with our machines. It's not too far-fetched to imagine that, someday, we'll be able to digitize and disembody our brain function. If we see ourselves as code, then perhaps, our code could be transferred to a new form, perhaps a robotic body or integrated with a new form of disembodied semi-consciousness. This type of disembodied snapshot of a mind may or may not still be us, but it could at least afford some level of limited immortality. As unappealing as this may sound to some, it will be considerably more appealing to many than being eaten by worms or scattered over the Himalayas.

Immortality in our current biological form will very likely prove impossible for us but, as Gilgamesh finally realized at the end of his epic quest to live forever, perhaps our immortality as a community comes from each of us contributing all we can to the greater good of society. Perhaps the best investment we can make in our immortality is to have a child, write a book, help save the environment, or contribute positively to our communities and cultures. To make more of that possible, why not do all we can to extend our healthy lives as much as our biology and technology can allow?

But as we shift our understanding of biology from something fixed, fated, and inevitable to being as readable, writable, and hackable as our information technology, we will need to stop fetishizing chance, rationalizing death to give meaning to life, or ascribing to spiritual mysticism those forces of mortality that we will increasingly comprehend.

The genetics, biotechnology, and longevity revolutions will challenge our current conceptions of what it means to be human beings. And we humans—with our frailties and superstitions, our primate brains, predatory instincts, and social systems honed over millions of years, and our limited biological capacities functionally merging with our seemingly limitless technologies—will need to figure out how to navigate the awesome ethical challenges just around the corner from where we now stand.

Chapter 8

The Ethics of Engineering Ourselves

Our history provides some wonderful examples of humans using technology to cure diseases, enhance our potential, explore the cosmos, and preserve our planet. We also have many examples of our using technology to kill and enslave each other, sow dissent, and destroy our surroundings. Our tools are agnostic. The variable is the values we individually and collectively apply when figuring out how we'll use them. The same will be true for the incredibly powerful tools of the genetic revolution. How we understand and apply these tools will significantly determine our future as a species. Each and all of our answers to the core questions of the genetic revolution will add up to the future trajectory of our species.

There are no easy answers to the critical questions of complexity, responsibility, diversity, and equity the genetics revolution will raise, but a better understanding of the ethical issues at stake is a critical first step.

We humans are each massively complex biological systems embedded in and constantly interacting with the even more complex ecosystems around us. Each of our genes performs multiple functions and interacts with other genes in ways we still don't fully understand. Even when we identify a gene that's doing something bad in one context, there's often a very real possibility the same gene is helping us in some other context.

We've seen that people who inherit a copy of the sickle cell mutation from each of their parents often suffer terribly from the disease. Those who inherit only one copy don't have sickle cell disease but can often possess a very significant natural resistance to malaria. Eliminating the sickle mutation from humans would end the suffering of sickle cell disease but potentially increase the suffering from malaria. Perhaps we could solve that problem by creating a more effective malaria drug, or we could use a gene drive to decimate the population of mosquitoes that transmit malaria.[1] But if we wiped out too many mosquitoes with a gene drive, we'd face the potential danger of crashing the ecosystems in which they live and doing even more damage to ourselves.

Each of these steps solves a particular problem but, for all we know, creates a new problem requiring an even greater intervention and so on down the line. It's like the old Peter, Paul and Mary song we used to sing at summer camp:

> *I know an old lady who swallowed a cow*
> *I don't know how she swallowed the cow*
> *She swallowed the cow to catch the goat*
> *She swallowed the goat to catch the dog*
> *She swallowed the dog to catch the cat*
> *She swallowed the cat to catch the bird*
> *She swallowed the bird to catch the spider*
> *That wriggled and jiggled and tickled inside her*
> *She swallowed the spider to catch the fly*
> *But I don't know why she swallowed that fly*
> *Perhaps she'll die.*[2]

Given that we still have so little understanding of how we function relative to the complexity of our biology, some ethicists argue, wouldn't we be better off not even attempting to play the role believers assign to god? As the conservative bioethicist Leon Kass wrote in the report of the George W. Bush–appointed U.S. Commission on Bioethics:

The dignity of being human, rooted in the dignity of life itself and flourishing in a manner seemingly issuing only in human pride, completes itself and stands tallest when we bow our heads and lift our hearts in recognition of powers greater than our own. The fullest dignity of the god-like animal is realized in its acknowledgement and celebration of the divine.[3]

Harvard philosopher Michael Sandel made a similar point in his *Atlantic* article and then book, *The Case Against Perfection.* "To believe that our talents and powers are wholly our own doing," he writes, "is to misunderstand our place in creation, to confuse our role with God's."[4]

This isn't just a philosophical position. We've already seen how early-stage embryos can be mosaics of different types of cells that make it difficult to determine whether a potentially harmful mutation will wind up a real problem for a future child or not. We've seen how even the most seemingly determinative of single-gene mutations don't always cause the disease they usually do. With so much at stake, why wouldn't we just take the safer route of trusting nature and our own biology—even with all of its bugs, shortcomings, and sometimes dangerous mutations—that's evolved over billions of years?

Because we are humans, that's why.

The moment our ancestors made tools, they were challenging the environment as they found it. The moment we started planting crops, clearing fields, and creating medicines to fight off the terrible, natural diseases afflicting us, we were giving the middle finger to nature as it was. Although we have good reasons to not want to cut down our forests, poison our air and oceans, and kill off the other species sharing our planet, our history with nature as we found it is a history of war. Nature conspired to kill us through hostile weather, predation, starvation, and disease, and we fought back with everything we had. We didn't live in harmony with nature; we balanced respect for nature with outright war against it.

We would all agree that ants are part of nature and so are anthills. Birds are part of nature and so are nests. Humans, too, are part of nature. What else could we be? If so, are not our shopping malls, nuclear bombs, and CRISPR edits also part of nature?

We would have no need to "play god" if the world we found ourselves inhabiting was less hostile. Darwin described nature as "clumsy, wasteful, blundering, low, and horribly cruel."[5] If god exists and is benign, it's fair to ask, why is god not playing god? If the world is in god's image as it is, why do we fight natural diseases like cancer at all? If it is not, why would we impose limits to what we might do to continue making the world a better and safer place for us? If god, for those who believe in the concept, didn't want us to live in caves fighting off disease, starvation, predators, and the elements, isn't everything from stone tools to fire to editing our genome part of doing our bit to complete god's unfinished business? But if, on the other hand, this god is agnostic about us, shouldn't we do what we can to defend and promote our species while protecting the ecosystems around us?

This leads to the logical conclusion not that we should never "play god" but that we must—but wisely and in balance with our surroundings. We will never have perfect knowledge of the genome just like we never had perfect knowledge of fire before we started harnessing it, cancer before we started treating it, or domesticated crops before we started eating them. In each case, we balanced the costs and benefits and moved forward with imperfect information and a hunger to learn and do more. The same is true with genetic technologies.

The bigger the step we consider taking to advance the genetic revolution, the more the potential costs will and should weigh on us. Continuing to conduct research is clearly the right thing to do. Making nonheritable genetic changes in people to help fight disease is also well on the path to widespread acceptance. Selecting and gene editing embryos will be bigger steps and rightly engender more conversation and scrutiny. But it would be folly to suggest that our species needs to wait for perfect knowledge to move forward. We can't and we won't. We are the hominins who climbed down from the trees and then conquered the world. We're filled with optimism and hubris. It's built into our operating system.

Opponents of any heritable genetic modifications warning of the "slippery slope," where each step leads to the next, are in this sense not wrong.[6] Each small step that scientists, doctors, or governments take in the direction of human genetic engineering will justify the next small step. If it's okay to incorporate another woman's donated mitochondria to help parents prevent a child from being born with mitochondrial disease, why couldn't heritable genetic changes be made to eliminate other terrible diseases? If it's okay to select an embryo who won't die young from Huntington's disease or doesn't have disrupted copies of the APOE4 gene that significantly increases its chances of getting early onset Alzheimer's, why couldn't we select one likely to live an extraordinarily long and healthy life? If we can mix and match genetic materials to make pigs with human hearts for transplant, why can't we mix and match genes from multiple humans to ensure our future children have the strongest human heart or other organ or capacity we can provide them?[7]

"Genetic illness and genetic wellness [are] not discrete neighboring countries; rather," Siddhartha Mukherjee beautifully writes in *The Gene*, "wellness and illness [are] continuous kingdoms, bounded by thin, often transparent, borders."[8] There is nothing inherently wrong with a slippery slope; we just need to be mindful of the direction in which we are sliding.

Decisions about whether or not to genetically select or alter preimplanted embryos are made by adults, but their real impact is on future children. Genetic conservatives, some of whom are political liberals, make the case that actively deciding the genetics of future children takes agency away from those children. But how much choice do children have about their mother's prenatal nutrition or stress levels during pregnancy, or whether they are sent, like many young South Korean kids, to cram schools from an early age to increase their chances of admission to the best universities and later success? While genetic conservatives argue we have a moral obligation to not alter future generations because they have no say in the matter, others contend just the opposite.

Oxford bioethicist Julian Savulescu is a key voice asserting that parents have a "moral obligation to create children with the best chance of the best life." He argues that "couples who decide to have a child have a significant moral reason to select the child who, given his or her genetic endowment, can be expected to enjoy the most well-being."[9] Oxford's Nick Bostrom arrives at the same conclusion when he suggests we deploy what he calls the "reversal test" when figuring out whether it is morally justified to make a given genetic alteration to a future human.[10]

Bostrom's argument is that if some people feel that making a change in one direction is bad—say, genetically engineering a person to eliminate a genetic disease—they would have to articulate why doing the opposite, that is, genetically engineering a person to add a genetic disease, would be justified, because this is in effect what they would be doing. Because no one could logically argue *for* genetically engineering humans to add diseases, this logic structure implicitly favors human enhancement.

Arguments like Savulescu's and Bostrom's have been attacked as morally repugnant by opponents who believe this justifies valuing one life more than another, denigrates disabled people, misunderstands what constitutes a "good life," conflates genetic therapies and enhancements, and commodifies future offspring. In his 2018 book *She Has Her Mother's Laugh*, science journalist Carl Zimmer warns against what he calls a "genetic essentialism" that reduces the complexity of humans and humanity to mere genetics.[11] All of these criticisms raise the specter of modern-day eugenics.[12]

The eugenics claim is not a spurious one. It hovers, in fact, over the entire prospect of human genetic engineering like a dark cloud.

X

The term *eugenics* combines the Greek roots for *good* and *birth*. Although coined in the nineteenth century, the concept of selective breeding and human population culling has a more ancient history. Infanticide was written into Roman law and practiced widely in the Roman Empire. "A father shall immediately put to death," Table IV of the Twelve Tables of Roman Law stated, "a son who is a monster, or has a form different from that of the human race."[13] In ancient Sparta, city elders inspected newborns to ensure that any who seemed particularly sickly would not survive. The German tribes, pre-Islamic Arabs, and ancient Japanese, Chinese, and Indians all practiced infanticide in one form or another.

The 1859 publication of Darwin's *The Origins of Species* didn't just get scientists thinking about how finches evolved in the Galapagos but about how human societies evolved more generally. Applying Darwin's principles of natural selection to human societies, Darwin's cousin and scientific polymath Sir Francis Galton theorized that human evolution would regress if societies prevented their weakest members from being selected out. In his influential books *Hereditary Talent and Character* (1885) and then *Hereditary Genius* (1889), he outlined how eugenics could be applied positively by encouraging the most capable people to reproduce with each other and negatively by discouraging people with what he considered disadvantageous traits from passing on their genes. These theories were embraced by mainstream scientific communities and championed by luminaries like Alexander Graham Bell, John Maynard Keynes, Woodrow Wilson, and Winston Churchill.

Although his work was partly in the spirit of the Victorian England times, Galton was then and even more now what we would call a racist. "The science of improving stock," he wrote, "takes cognizance of all the influences that tend in however remote degree to give the more suitable races or strains of blood a better chance of prevailing speedily over the less suitable than they otherwise would have had."[14] In 1909, Galton and his colleagues established the journal *Eugenics Review*, which argued in its first edition that nations should compete with each other in "race-betterment" and that the number of people in with "pre-natal conditions" in hospitals and asylums should be "reduced to a minimum" through sterilization and selective breeding.[15]

Galton's theories gained increasing prominence internationally, particularly in the New World. Although eugenics would later accrue sinister connotations, many of the early adopters of eugenic theories were American progressives who believed science could be used to guide social policies and create a better society for all. "We can intelligently mold and guide the evolution in which we take part," progressive theologian Walter Rauschenbusch wrote. "God," Johns Hopkins economic professor Richard Ely asserted, "works through the state." Many American progressives embraced eugenics as a way of making society better by preventing those considered "unfit" and "defective" from being born. "We know enough about eugenics so that if that knowledge were applied, the defective classes would disappear within a decade," University of Wisconsin president Charles Van Hise opined.[16]

In the United States, the "science" of eugenics became intertwined with disturbing ideas about race. Speaking to the 1923 Second International Congress of Eugenics, President Henry Osborn of New York's American Museum of Natural History argued that scientists should:

> ascertain through observation and experiment what each race is best fitted to accomplish... If the Negro fails in government, he may become a fine agriculturist or a fine mechanic... The right of the state to safeguard the character and integrity of the race or races on which its future depends is, to my mind, as incontestable as the right of the state to safeguard the health and morals of its peoples. As science has enlightened government in the prevention and spread of disease, it must also enlighten government in the prevention of the spread and multiplication of worthless members of society, the spread of feeblemindedness, of idiocy, and of all moral and intellectual as well as physical diseases.[17]

Major research institutes like Cold Spring Harbor, funded by the likes of the Rockefeller Foundation, the Carnegie Institution of Washington, and the Kellogg Race Betterment Foundation, provided a scientific underpinning for a progressive eugenics movement growing in popularity as a genetic determinism swept the country. The American Association for the Advancement of Science put its full weight behind the eugenics movement through its trend-setting publication, Science.[18] If Mendel showed there were genes for specific traits, the thinking went, it was only a matter of time before the gene dictating every significant human trait would be found. Ideas like these moved quickly into state policies.

Indiana in 1907 became the first U.S. state to pass a eugenics law making sterilization mandatory for certain types of people in state custody. Thirty different states and Puerto Rico soon followed with laws of their own. In the first half of the twentieth century, approximately sixty thousand Americans, mostly patients in mental institutions and criminals, were sterilized without their acquiescence. Roughly a third of all Puerto Rican women were sterilized after providing only the flimsiest consent.[19] These laws were not entirely uncontroversial, and many were challenged in courts. But the U.S. Supreme Court ruled in its now infamous 1927 *Buck v. Bell* decision, that eugenics laws were constitutional. "Three generations of imbeciles," progressive Supreme Court justice Oliver Wendell Holmes disgracefully wrote in the decision, "are enough."[20]

As the eugenics movement played out in the United States, another group of Europeans was watching closely. Nazism was, in many ways, a perverted heir of Darwinism. German scientists and doctors embraced Galton's eugenic theories from the beginning. In 1905, the Society for Racial Hygiene was established in Berlin with the express goal of promoting Nordic racial "purity" through sterilization and selective breeding. An Institute for Hereditary Biology and Racial Hygiene was soon opened in Frankfurt by a leading German eugenicist, Otmar Freiherr von Verschuer.

Eugenic theories and U.S. efforts to implement them through state action were also very much on Adolf Hitler's mind as he wrote his ominous 1925 manifesto, *Mein Kampf*, in Landsberg prison. "The stronger must dominate and not mate with the weaker," he wrote:

> Only the born weakling can look upon this principle as cruel, and if he does so it is merely because he is of a feebler nature and narrower mind; for if such a law did not direct the process of evolution then the higher development of organic life would not be conceivable at all... Since the inferior

always outnumber the superior, the former would always
increase more rapidly if they possessed the same capacities
for survival and for the procreation of their kind; and the
final consequence would be that the best in quality would be
forced to recede into the background. Therefore a corrective
measure in favor of the better quality must intervene...for
here a new and rigorous selection takes place, according to
strength and health.[21]

One of the first laws passed by the Nazis after taking power in 1933
was the Law for the Prevention of Hereditary Defective Offspring, with
language based partly on the eugenic sterilization law of California. Genetic
health courts were established across Nazi Germany in which two doctors
and a lawyer helped determine each case of who should be sterilized.

Over the next four years, the Nazis forcibly sterilized an estimated four
hundred thousand Germans. But simply sterilizing those with disabilities
was not enough for the Nazis to realize their eugenic dreams. In 1939, they
launched a secret operation to kill disabled newborns and children under
the age of three. This program was then quickly expanded to include older
children and then adults with disabilities considered to have *lebensunwertes
leben*, or lives unworthy of life.

Making clear the conceptual origins of these actions lay in scientifically and medically legitimated eugenics, medical professionals oversaw
the murder of an ever-widening group of undesirables in "gassing instal-
lations" around the country. This model then expanded from euthanizing
the disabled and people with psychiatric conditions to criminals and to
those considered to be racial inferiors, including Jews and Roma, as well as
homosexuals. It was not by accident that Joseph Mengele, the doctor who
decided who would be sent to the gas chambers at Auschwitz, was a former
star student of von Verschuer at the Frankfurt Institute for Hereditary
Biology and Racial Hygiene.

By the mid-1930s, the American scientific community was pulling away
from eugenics. In 1935, the Carnegie Institution concluded the science of

eugenics was not valid and withdrew its funding for the Eugenics Records Office at Cold Spring Harbor. Reports of Nazi atrocities amplified by the 1945–46 Nuremberg trials put the nail in the coffin of the eugenics movement in the West. Although eugenics laws were finally scrapped from the books only in the 1960s in the United States and the 1970s in Canada and Sweden, very few people were forcibly sterilized after the war.

But as new technologies more recently began to revolutionize the human reproduction process and create new tools for assessing, selecting, or genetically engineering preimplanted embryos, many critics raised the specter of eugenics. In his influential 2003 presidential address to the American College of Medical Genetics, and later in a published article, University of California, San Francisco, pediatrician Charles Epstein provocatively asked, "Is modern genetics the new eugenics?" He answered that it had the potential to be, if the scientific community was not self-aware and careful.[22]

Harvard's Michael Sandel made a similar point in *The Case Against Perfection*:

> Genetic manipulation seems somehow worse—more intrusive, more sinister—than other ways of enhancing performance and seeking success… This draws it disturbingly close to eugenics… Was the old eugenics objectionable only insofar as it was coercive? Or is there something inherently wrong with the resolve to deliberately design our progeny's traits… The problem with eugenics and genetic engineering is that they represent a one-sided triumph of willfulness over giftedness, of dominion over reverence, of molding over beholding.[23]

Leading bioethicist Arthur Caplan argued:

> Renegade scientists and totalitarian loonies are not the folks most likely to abuse genetic engineering. You and I are—not

because we are bad but because we want to do good. In a world dominated by competition, parents understandably want to give their kids every advantage... The most likely way for eugenics to enter into our lives is through the front door as nervous parents...will fall over one another to be first to give Junior a better set of genes.[24]

The parallels between the ugly eugenics of the late nineteenth century and the first half of the twentieth and what's beginning to happen today are not insignificant. In both cases, a science at an early stage of development and with sometimes uncertain accuracy was or is being used to make big decisions—forced sterilization of the "feeble-minded" in the old days, not selecting a given embryo for implantation or terminating a pregnancy based on genetic indications today. In both cases, scientists and government officials seek to balance individual reproductive liberty with broader societal goals. In both cases, future potential children lose the opportunity to be born. In both cases, societies and individuals make culturally biased but irrevocable decisions about which lives are worth living and which are not. These parallels offer us a powerful warning.

But if we collectively paint all human genetic engineering with the brush of Nazi eugenics, we would kill the incredible potential of genetics technologies to help us live healthier lives. "If cannibalism is our greatest taboo," Oxford philosopher Richard Dawkins wrote, "positive eugenics... is a candidate for the second... In our time, the word has a chilling ring. If a policy is described as 'eugenic,' that is enough for most people to rule it out at once."[25] Bioethicist Diane Paul writes that the term *eugenics* is "wielded like a club. To label a policy 'eugenics' is to say, in effect, that it is not just bad but beyond the pale."[26] That there probably is an element of eugenics in decisions being made today on the future of human genetic engineering should push us to be careful and driven by positive values, but the specter of past abuses should not be a death sentence for this potentially life-affirming technology or the people it could help.

Unlike the earlier eugenics movement, today's models of prenatal testing and embryo selection are not state controlled, coercive, racist, or discriminatory by standard uses of those terms. "Modern eugenic aspirations are not about the draconian top-down measures promoted by the Nazis and their ilk," journalist Jon Entine writes. "Instead of being driven by a desire to 'improve' the species, new eugenics is driven by our personal desire to be as healthy, intelligent and fit as possible—and for the opportunity of our children to be so as well. And that's not something that should be restricted lightly."[27] Entine's point can be debated, but for our and our species' sake, we must be having the debate.

A basic tenet of liberal societies is that wherever possible the individual should be protected from the excessive power of the state. One expression of this philosophy in many societies is the protection of a woman's right to make her own reproductive decisions on issues like contraception and abortion. "Reproductive freedom," New Zealand philosopher Nicholas Agar writes in his book *Liberal Eugenics: In Defense of Human Enhancement*, "encompasses the choice of whether or not to reproduce, with whom to reproduce, when to reproduce, and how many times to reproduce... Liberal eugenics adds the choice of certain of your children's characteristics to that list of freedoms."

Agar contrasts liberal eugenics with what he calls "authoritarian eugenics," the idea that the state should determine what constitutes a good life. "While old-fashioned authoritarian eugenicists sought to produce citizens out of a single centrally designed mold," Agar writes, "the distinguishing mark of the new liberal eugenics is state neutrality. Access to information about the full range of genetic therapy will allow prospective parents to look to their own values in selecting improvements for future children. Authoritarian eugenicists would do away with ordinary procreative freedoms. Liberals instead propose radical extensions of them."

As opposed to decisions made by governments in authoritarian systems based on race and class, liberal eugenics systems would focus on the individual. In the society Agar envisions, "parents' particular conceptions of the good life would guide them in their selection of enhancements for their children,"[28] but these parental choices would need to be balanced by consideration of the broader societal impact of their individual decisions.

Parental choice, however, is an important principle but not an unlimited one. The United States, for example, has upheld the right of Amish parents to separate their children from modernity and of Hasidic Jews to raise their children speaking only Yiddish and not learning English. Many U.S. states offer less latitude for pregnant women who poison their fetuses through drug and alcohol use while pregnant or for Christian Science parents who deny their children urgent medical care. It's also probably impossible to think of individual parental choices outside of the social and political context.

Given that selecting embryos to avoid disease is already permitted in most jurisdictions around the world, it's unlikely other reproductive decisions intended to enhance the health and well-being of future children will be forever banned in most countries. When this happens, the essential question for parents will move from whether to select and ultimately manipulate embryos but how much. Some will call this eugenics, but the connotations will change.

In his provocative 2017 *Los Angeles Times* editorial "Is There Such a Thing as Good Eugenics?" Adam Cohen asserts, "Twentieth century eugenics has rightly been called a 'war on the weak... [t]wenty-first century eugenics...can be a war for the weak."[29] This point is necessarily controversial, but just as genetically engineering our future children comes with a potential cost, so does the alternative.

It's not that hard to imagine future scenarios when humans would need to genetically alter ourselves in order to survive a rapid change in our environment resulting from global warming or intense cooling following a nuclear war or asteroid strike, a runaway deadly virus, or some kind of other future challenge we can't today predict. Genetic engineering, in other words, could easily shift from being a health or lifestyle choice to becoming an imperative for survival. Preparing responsibly for these potential future dangers may well require we begin developing the underlying technologies today, while we still have time.

Thinking about genetic choice in the context of imagined future scenarios is, in many ways, abstract. But potentially helping a child live a healthier, longer life is anything but. Every time a person dies, a lifetime of knowledge and relationships dissolves. We live on in the hearts of our loved ones, the books we write, and the plastic bags we've thrown away, but what would it mean if people lived a few extra healthy years because they were genetically selected or engineered to make that possible? How many more inventions could be invented, poems written, ideas shared, and life lessons passed on? What would we as individuals and as a society be willing to pay, what values might we be willing to compromise, to make that possible? What risks would we individually and collectively be willing to take on? Our answers to these questions will both propel us forward and present us with some monumental ethical challenges.

X

Some people today see diversity as a way to redress historical wrongs committed against minority populations. Others see it as a way of making sure our universities, corporations, governments, and other institutions benefit from a wide range of perspectives. All of these understandings miss an even more essential point: diversity has been and remains the sole survival strategy of our species.

Random mutation, one of the two pillars of Darwinian evolution, is just another name for the precursor of diversity. Diversity allowed our single-cell ancestors to morph into what we are today. Diversity was what both ensured those species without helpful adaptations would die off and those benefiting from adaptations more suitable to the changing world around them would thrive. Evolution sometimes feels like an inexorable progression from better to worse, from the knuckle-walking monkey to the upright man, but this is not the case. Evolution is directionless and agnostic about its winners and losers. Conditions are always changing, so evolving creatures generally have no sense they are evolving and no possible way of knowing which traits will be selected for, or against, in the context of future conditions they cannot imagine.

If given a choice, monkeys in trees would probably choose to make themselves better climbers. Our tree-living ancestors must have looked at that first lunatic who climbed down as some type of freak, but the adaptations associated with savannah living conferred advantages that ensured the associated genetic mutations persisted. Most of our ancestors were not the first early humans to leave Africa, only the first early humans to leave Africa and ultimately survive. They lasted presumably because they evolved a set of traits that gave them a better chance of survival within their new context. Had we asked our ancestors at the point of leaving Africa what traits they would need, they would have had no way of knowing. Genetic diversity conferred a range of traits, some of which made survival possible for a small subset of people.

Diversity, therefore, is conferred upon us, but we would, by definition, be hard-pressed to overcome our biases of the present moment. Left to our own devices, we humans are easily swayed by what we see around us. A recent Australian study that reviewed the records of 154 women's selections of donors at a Queensland sperm bank found that women preferred well-educated, introverted, and analytical donors over those who were less educated, extroverted, and less methodical.[30] If given the choice, it's a good guess that most parents would want smarter, taller, better looking, and more compassionate kids because these traits are valued in today's world.[31] Each of these traits, however, reflects our specific cultural bias.

The closer we move toward sensitive and culturally significant issues like race, sexual orientation, and intelligence, the greater the danger we might collectively be swayed by pernicious social biases. Very often concerned people ask me in the question-and-answer session following my talks whether embryo selection or gene editing could be used for parents to not have a gay child or to make sure their child is light-skinned. The answer is maybe.

Although there is no single "gay gene," the irrefutable case has been made through twin and genome-wide association studies that there is at least a significant genetic foundation to homosexuality.[32] It would be impossible to today determine which among the fifteen or so embryos being considered for implantation in the mother would have the greatest likelihood of being gay, but this may not always be the case.

The same is true for skin color. A 2018 study by researchers at the University of Pennsylvania identified eight genes that significantly influence a person's skin pigmentation.[33] A single mutation of the SLC24A5 gene accounts for most of what we call white skin. It is likely that parents with different skin tones selecting from among their embryos will in the not-too-distant future be able to choose one that has a lighter or darker complexion. In the longer term, it will probably be possible to genetically edit a preimplanted embryo or even perform gene therapy on adults to change their color. The usual human colors wouldn't necessarily be the only options. Japanese scientists have already used CRISPR to change the colors of flowers from purple to white by disrupting a single gene.[34] It is only a matter of time until a wide range of color-changing options will be available to humans, should we choose to use them. Choices like these that could reduce human diversity would not only have significant and negative social implications but also potentially expose us to yet-unknown risks.

Over the past eight thousand years our ancestors domesticated corn to increase yield, but because this modified corn became so genetically homogenous its susceptibility to blight increased. Similarly, if enough people around the world start making individual decisions about their future children based on whatever cultural biases, we run the collective risk of weaponizing our biases and making our species more genetically uniform and less able to withstand some type of natural virus, germ warfare, or other yet unknown challenge in the future. If we don't recognize the evolutionary benefits of our diversity in its many facets, we run the risk of harming ourselves, even in the name of our individual and collective efforts to help ourselves.

In 2017, a U.S. military contractor invited me to a conference that brought together futurists and military leaders to explore what impact the genetics revolution might have on the future of warfare and how the U.S.

military might prepare. After dividing into breakout sessions designed to generate different scenarios for what might happen, each group came back to share their thoughts. The first team outlined how future soldiers could be genetically engineered to be great at pattern recognition.

"Why don't we just recruit high-functioning autistic people into the military who are already far better at solving certain types of cognitive problems than non-autistics?" I piped in, trying not to be snarky. "Why genetically engineer people with great pattern recognition when we already have them?"[35]

These types of questions must broadly be asked as we enter the age of human genetic engineering. Why breed people for superior IQ when we could significantly enhance our society's collective intellectual and problem-solving capability by providing real opportunity and better schools to the least advantaged among us? Why use embryo selection to reduce the rates of bipolar disorder when some of the world's greatest artists are great because their brains function differently?[36]

These are tough questions for which there are no easy answers, but the idea of parents selecting embryos less likely to become gay adults or biracial parents choosing or gene editing kids to be one color or another is not only morally repulsive but also a potential blow to the beneficial diversity of our species. Because these could well be potential choices in the future, we must redouble our efforts today to find better and new ways to celebrate and invest in our diversity and ensure our common genetic data pool is diverse to prevent the genetic revolution from ultimately becoming a horror show. Diversity is our species' greatest asset should we choose to embrace it. Reducing our diversity, even with the best of intentions, could become our Achilles' heel.

On the other hand, it would also be a mistake to over-fetishize every aspect of genetic diversity. A few genes can save us, others can kill us, and most everything else sits somewhere in the middle being experienced differently in different contexts. To suggest we should accept the terrible genetic diseases that kill our children out of respect for genetic diversity or that we should allow parents or scientists to engineer children with debilitating diseases to enhance diversity would be preposterous. And as much as genetic engineering might be used to limit genetic diversity, it could also conceivably be used to enhance it. Either way, to realize the greatest upside of the genetic revolution we will have to articulate, celebrate, and affirm in our individual and collective choices the value of diversity in the coming genetic age in a far more profound manner than we do today.

Like diversity, issues of equity will also need to be front and center in our thinking.

Ж

All technology has an uneven adoption curve.

There's always a first person or a first group to have access to a particular advantageous technology. Scientists are divided about whether our Homo sapiens ancestors out-competed, out-hunted, out-survived, and/or out-humped our Neanderthal cousins to extinction, but whatever the case we used our tools, social strategies, sex organs, and brains to devastating impact. Tens of thousands of years later, the Mongols built the largest geographic empire in history, marrying the military advantage that stirrups gave their cavalry to a ruthless warrior culture and new ideas for social organization. Although China invented gunpowder and the compass, these technologies were fully weaponized into guns and warships by the Europeans, who used these capabilities to colonize the rest of the world. In each of these cases, a relatively small technological advantage was levered into decades or centuries of domination or, in the case of the Neanderthals, extinction.

There are many ways the unequal distribution of genetic technologies could lead to these kinds of frightening results. If only wealthy and otherwise advantaged people can select or genetically engineer their children to have certain useful traits, their children could come to dominate societies because of their real or perceived capabilities. Employers might not want to take the risk of hiring someone who was not enhanced if the odds of an enhanced person doing the same job better, by whatever metric, were statistically higher or even if there was only a false perception this was the case. If enhancements are allowed and unequal access continues over time, each generation of an advantaged family could become more genetically enhanced than their disadvantaged peers until the difference between two groups becomes unbridgeable.

But despite the critical importance of striving for greater levels of equality, absolute equality should not be a goal. The popular 1997 film *Gattaca*, in which an unenhanced young man wanting to become an astronaut is blocked because of his genetic profile, explored this idea. The protagonist eventually makes it to space through grit, wit, and wile. In real life, would a society really want a genetically nonenhanced person traveling to space if someone else was genetically optimized to be more resistant to radiation and maintain bone density better in a zero-gravity environment?

Many technologies start out being used by a few elites before later reaching a wider audience. Meeting with female villagers participating in a microcredit program during my 2012 visit to Bangladesh, I was incredibly impressed by what all the women receiving the loans were doing to start small businesses and take care of their families but saddened that they seemed to have little chance to do more. With my iPhone connecting me to the universal library of the internet, I felt the already huge advantage of my privileged American birth widening. How could these poor people ever afford the expensive technological marvel I held in my pocket? I wondered. Today, Bangladeshi villagers can get a new smartphone for an affordable $60, and usage levels are skyrocketing. If we had demanded equal access to smartphones for everyone from the get-go, the smartphone industry would never have grown quickly enough to drive down prices to where these phones became accessible to poor people around the world.

We've already explored how governments and insurance companies will be incentivized to promote embryo screening and then gene editing to eliminate genetic diseases and avoid the cost of providing a lifetime of care to those who would otherwise be born with those diseases or develop them later in life. But saying that companies and governments will be incentivized to make embryo screening and genetic engineering more widely available does not mean they will do it effectively or at all, because even the inefficient status quo always has its protectors. On top of that, some parents with means will want to select and enhance their future children more aggressively despite whether their governments or insurance companies will pay for it. There are no easy answers, but it's fair to ask whether preventing the first adopters from genetically enhancing their children would be the same as preventing the early adopters of smartphones and supercomputers from leveraging the advantages those technologies provided.

And while many people fear a dystopian future of genetic determinism, the case for genetic identification should not be rejected out of hand. Mozart grew up in the Hapsburg court, but how many Mozarts are today languishing in Syrian refugee camps? Will we always think it wrong for music programs to know which of their young applicants are genetically predisposed to having perfect pitch?[37] Will we oppose the genetic screening of the world's most disadvantaged communities to identify children with tremendous genetic potential and give them a better shot at realizing that potential or of all children to assess what teaching styles may best correspond to their learning abilities?

No one wants to live in a society where people are pegged at birth for certain roles and never have the opportunity to find their passions and show what they can achieve, and we must do all we can to provide equal opportunity to everyone. But providing extra opportunities for people with remarkable genetic potential in one area or another may come over time to be seen as a service to disadvantaged communities, a boon to national competitiveness, and the good and right thing to do. It could well be that some of us will want, over the coming decades and centuries, for our children to have enhanced abilities and traits.

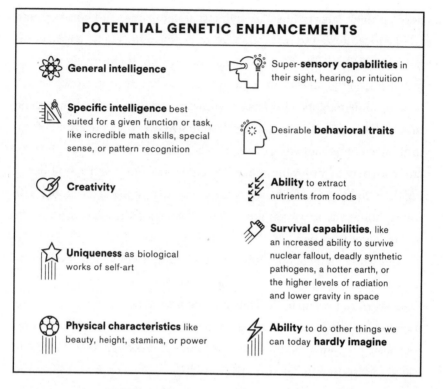

POTENTIAL GENETIC ENHANCEMENTS

General intelligence

Specific intelligence best suited for a given function or task, like incredible math skills, special sense, or pattern recognition

Creativity

Uniqueness as biological works of self-art

Physical characteristics like beauty, height, stamina, or power

Super-**sensory capabilities** in their sight, hearing, or intuition

Desirable **behavioral traits**

Ability to extract nutrients from foods

Survival capabilities, like an increased ability to survive nuclear fallout, deadly synthetic pathogens, a hotter earth, or the higher levels of radiation and lower gravity in space

Ability to do other things we can today **hardly imagine**

Even if enhancements like these were not evenly distributed, a compelling case can be made that a small number of enhanced people, if motivated by positive values, could make tremendous contributions to various fields like science, philosophy, art, or politics that could make the world better for everyone.

As we co-evolve with our technology, it may be that we'll need to generate a group of brilliant coders to extend the human role in our human-machine interface. It may be that human creativity and human qualities like empathy become so valued in an AI-defined world that we will start an arms race with each other to genetically engineer more empathic and creative children. As frightening as this might seem to some, restricting the genetic enhancement capacities we may need to maintain our species' position in a world of artificial superintelligence might end up being like limiting horse and buggy speeds at the dawn of the automotive age.

Accepting that identifying genetic predispositions for a certain function could be justified, however, does not mean that we should passively accept a future where perceived capacities are determined by genetics or where the gap between genetically enhanced and unenhanced people continually grows.

X

After every talk I give on the future of human genetic engineering, someone asks about the future dangers of genetically engineered inequality. My answer is always the same. Genetic inequality should be a very real concern for our future. But if we are concerned about inequality at some distant point in the future, we should start living our values of equality today. Addressing the current difference between the average person reading this book and the average resident of the Central African Republic would be a good place to start.

Because of chronic civil war, 76 percent of the Central African Republic population lives in poverty, a quarter is displaced, half are food insecure, and 40 percent of young children are experiencing stunted growth. Widespread maternal malnutrition suggests that the cognitive function of these children is very likely, on average, significantly lower than that of average children born into more advantaged environments.[38] In comparison to their Central African Republic counterparts, therefore, advantaged children from other countries are already genetically enhanced. If we care about equality—and genetic equality—as we should, advocating for it in our divided world today will model our best values for tomorrow, when genetically engineered inequality becomes a possibility.

As we aspire to live these values, however, we must also remember that some genetic inequality is part of being human, a central aspect of our diversity. The truest nightmare scenario for our species is not genetic inequality but complete genetic equality. At the same time, inequality run amok would become a terrible nightmare. The Goldilocks solution to this challenge may be that we need to find a balance between the excesses of too much and too little genetic equality. But members of a hubristic species like ours must at least consider challenges to such a seemingly reasonable hypothesis.

In his introduction of the idea of the *Übermensch*, or superman, the German philosopher Friedrich Nietzsche recognized what it meant that humans had not plopped on the earth fully made, as the Bible suggests, but were instead products of an ongoing evolutionary process with our current form only one step along the way. This transitory version of us, according to Nietzsche, was something to be overcome. "All beings so far," he asked, "have created something beyond themselves; and do you want to be the ebb of this great flood and even go back to the beasts rather than overcome man?"[39]

Julian Huxley, a zoologist and leader of the British Eugenics Society (and brother of *Brave New World* author Aldous), was an avowed eugenicist in the years before the Second World War who supported the voluntary sterilization of people with "mental defects" and restrictions on immigration into the United Kingdom. Even after Nazism drove a nail into the heart of the eugenics movement, Huxley struggled for years to define a more modern eugenics, based on principles of "scientific humanism."[40] In 1957, the same year that the Soviets launched Sputnik and ushered the world into a new era of big science, he wrote in his essay "Transhumanism" that:

> The human species can, if it wishes, transcend itself—not just sporadically, an individual here in one way, an individual there in another way—but in its entirety, as humanity. We need a name for this new belief. Perhaps transhumanism will serve: man remaining man, but transcending himself, by realizing new possibilities of and for his human nature…the human species [can] be on the threshold of a new kind of existence, as different from ours as ours is from that of Peking man. It will at last be consciously fulfilling its real destiny.[41]

In the years since, the transhumanist movement has grown from an intellectually aligned group of thinkers and technologists into a global movement with its own international association, manifesto, and even candidates for political office. Championed by brilliant thinkers like Hans Moravec, Ray Kurzweil, and Nick Bostrom, transhumanists, according to the 1998 *Transhumanist Declaration*, "believe that humanity's potential is still mostly unrealized." This potential can be broadened "by overcoming aging, cognitive shortcomings, involuntary suffering, and our confinement to planet Earth." Individual humans should be afforded "wide personal choice over how they enable their lives," including the "use of techniques that may be developed to assist memory, concentration, and mental energy; life extension therapies; reproductive choice technologies; cryonics procedures; and many other possible human modification and enhancement technologies."[42] Like eugenics and most religions before it, transhumanism imagines a world where humans transcend the limitations of their current biology.

But while Nietzsche's imagining of the *Übermensch* gained sinister implications,[43] the transhumanists and others like them are imagining an *Über*-intelligence combining human and machine capabilities into a seamless web of evolutionary revolution. These ideas are at the same time frightening and tantalizingly appealing.

The entire history of human existence has been marked by our incessant struggle to increase our chances of survival by becoming better at securing calories, protecting ourselves from the elements, and procreating. With every three steps forward we've made as a species, we've fallen two steps back as lifetimes of learning succumb to the aging decay of our brains and the destruction of death. If genetic technologies help us live healthier and longer, retain knowledge, and do many other things better, or even just make us feel we have the capacity to fight back against the caprice of our own biology, the magnetic pull of using these technologies will prove collectively irresistible.

But the intellectual connectivity between eugenics and transhumanism faces all of us as a warning. Saying that we will and probably should do something does not mean that there should be no limits. Even the most liberal societies have laws regulating what people can and can't do in their intimate lives and beyond. The future of genetic engineering is in many ways the future of humanity. To enable our future, it must be embraced. To save us from ourselves, however, it must be regulated.

The challenge we face today, however, is that while the science is advancing exponentially, public understanding of it is only increasing linearly. The regulatory structures needed to find the right balance between scientific progress and ethical constraint are only inching forward glacially. And all of this is happening in a world where significant cultural and societal differences, driven by ever-greater levels of competition within and between societies, make it ever more difficult to find common ground.

Chapter 9

We Contain Multitudes

We humans aren't just a genetically diverse species. We are also culturally diverse. It's a big part of what makes our world so interesting and our relationships so rewarding.

This diversity of opinion and approach is a great asset for our species but also sometimes comes at a cost. We approach a range of issues related in one way or another to the future of human genetic engineering—including how we treat the environment, how we grow our food, and how we think about when life begins—very differently. When differences over issues like these became too great, we have argued, we have competed politically to gain an upper hand, and sometimes we have fought to prove our righteousness. Exploring how we handled these past debates on issues adjacent to human genetic engineering gives us both an indication of what we might expect as the genetic revolution plays out and a warning about just how difficult it will be to chart a common path forward.

Genetic engineering involves altering nature in ways our ancestors could hardly have imagined. Our forebears swimming in the oceans, crawling onto the land, and swinging in the trees didn't have the capacity to massively disrupt the environment. We were just one among many species struggling to survive in the hostile and dangerous world around us, facing repeated bouts of near starvation and ravaged by diseases, and fighting off predators trying to kill us.

But as soon as we could, we humans fought back hard against the vagaries of nature. We developed weapons to kill other animals and occasionally each other, agriculture to secure a stable source of calories, and medicines to fight against the afflictions of nature. Some human communities were more rapacious than others, but all molded the world to our needs.

Even hunter-gatherer societies with traditions revering the natural forces around them were likely responsible for mass extinctions of other species. After the ancestors of the Native American Indians crossed the Bering Strait thirteen thousand years ago, for example, huge numbers of species in the new world—including mammoths, mastodons, dire wolves, giant beavers, and camelops—quickly went extinct.[1] The same destruction of the natural habitat happened after the first humans arrived in Hawaii, New Zealand, Easter Island, and many other places.

Early Western theologians referenced the Old Testament to justify human domination of nature. "Let us make man in our image," god says in Genesis 1:26, "and let them have dominion over the fish of the sea, and over the fowl of the air, and over the cattle, and over all the earth, and over every creeping thing that creepeth upon the earth."[2] After using the philosophy of dominion to justify conquering the natural world and destroying their environment at home, European societies used their new scientific and industrial capabilities to brutally colonize and ruthlessly exploit the rest of the world. In the second half of the twentieth century, majority European sensibilities shifted, and Europeans became the world's leading environmentalists and champions of global efforts to slow climate change.

European settlers in the United States cut down vast forests to build plantations, wiped out large ecosystems when covering the land with farms, and hunted animals like the bison to near-extinction for sport. But then preservationists like John Muir inspired Teddy Roosevelt to create America's first national parks, and environmental consciousness grew. In the three remarkable years between 1969 and 1971, Friends of the Earth, the Natural Resources Defense Council, Greenpeace, and the U.S.

Environmental Protection Agency (EPA) were created and the first Earth Day was proclaimed.*

China has similarly undergone a series of about-faces in popular and governmental perceptions of nature. Traditional Chinese culture valued a sense of harmony between humans and nature. The unity of man and nature, *tian ten he yi*, is prominent in major Chinese schools of thought like Confucianism and Taoism. "If close nets are not allowed to enter the pools and ponds," the great fourth-century BC philosopher Mencius wrote, "the fishes and turtles will be more than can be consumed. If the axes and bills enter the hills and forests only at the proper time, the wood will be more than can be used."[3] As with other cultures, this harmony was an ideal not always realized.

After Mao and the Chinese Communists seized power in 1949, this nature-respecting traditional philosophy was turned on its head with a vengeance. In an effort to speed up China's modernization, Mao launched the Great Leap Forward in 1958. To build an agricultural surplus to support industrialization, peasant farmers were forced into communes and told to plant up to ten times more seeds in their fields than they had previously done. The crops died from over-density. Because sparrows were eating some of the crops that remained, Mao and party leaders called on children to search out and destroy sparrow nests and on farmers to bang pots and pans together to drive the sparrows to exhaustion. After millions of sparrows were killed, the insect populations predictably exploded, further devastating agricultural yields. Even in this downward spiral, peasant farmers were ordered to build backyard furnaces to melt metals that could be used to speed up industrialization. Trees around the country were cut down to make wood for these useless furnaces, decimating China's forests. When drought arrived, Mao's China was wholly unprepared. Between 1958 and 1962, an estimated forty-five million Chinese

---

*   Under the Donald Trump presidency, this burgeoning environmental progress slipped away, and the U.S. government pulled out of its previous commitments to address climate change, scaled down national parks, and gutted the EPA.

people died, the greatest man-made famine in history on the back of Mao's great ecocide.[4]

China's destruction of its environment continued after Mao, even as the country implemented smarter industrialization policies. In 1978, newly installed premier Deng Xiaoping began a process of opening China's economy that paved the way for the country's rapid growth. Led mostly by engineers, the Chinese Communist Party behaved as if all of life could be engineered. It adopted a one-child policy to engineer the population, massive works projects like the Three Gorges Dam and the South-to-North Water Diversion Project to engineer the environment, and industrial policies to engineer growth with little regard for pollution, safety, or worker rights. As China became the fastest growing economy in history over the next four decades, the grow-at-all-costs philosophy led to the contamination of the vast majority of its waterways, the poisoning of its air, and the transformation of previously verdant lands into deserts.[5] But facing the threat of revolt from newly empowered middle class choking on China's air and being poisoned by its food, recent Chinese leaders like Xi Jinping have made cleaning up China's environment a priority.

All of these twists and turns in national approaches to environmental protection make clear that the national and communal ideologies that will guide our future actions on genetically altering ourselves will not be uniform within or especially between societies.

At its core, the coming debate over human genetic engineering will be a debate about how far we as a species—made up of many diverse groups with very different views on whether and how much to alter the biology bequeathed to us by evolution—will go. These types of distinct philosophies about our relation to nature have played out very differently in various societies in our contentious debates over genetically modified crops.

✕

Our ancestors were selectively breeding plants long before Gregor Mendel figured out the mechanism governing the transfer of traits across generations. Understanding the rules of genetics made researchers even better able to breed all sorts of organisms with particular genetic traits. By understanding more of what genes are and do, scientists were able to go one step further by transferring genetic material from one organism to another.

In 1973, Stanford Medical School graduate student Stanley Cohen and his professor Herbert Boyer transferred a gene that provided antibiotic resistance from one strain of bacteria to another that lacked this resistance. When the second bacteria became antibiotic-resistant, the new era of genetically modified organisms, GMOs, was born.* GMOs are plants, animals, and other organisms whose genetic material has been changed by humans to a form that does not generally occur on its own in nature, particularly by transferring genes between species.

Soon after, a microbiologist working for General Electric filed a patent for a genetically engineered bacteria designed to break down crude oil, potentially very useful in addressing oil spills. GE's patent application was rejected because the U.S. Patent Office rules clearly indicated that living things could not be patented. But after GE appealed, the U.S. Supreme Court shocked the world in its 1980 ruling that "a live, human-made microorganism is patentable subject matter."[6] If life could be patented, the race was on to build the portfolio.

But as new companies like Genentech and Amgen rapidly built their businesses, others began worrying about the potential dangers of what scientists began calling "recombinant DNA," the process of combining genetic materials from multiple organisms to create genetic sequences not otherwise found in nature. A prominent group of scientists called on the director of the U.S. National Institutes of Health to establish a special committee to evaluate the potential biological and ecological hazards of this new technology and ways of preventing the unintended spread of recombinant

---

* Genetically modified organisms are ones whose genes have been altered in ways that don't normally occur through mating. Transgenic organisms are ones where the genetic modification involved adding genetic material from a different organism.

DNA molecules across the human, animal, and crop populations.[7] In 1975, some of America's leading scientists got together with ethicists, lawyers, and government officials at the Asilomar Conference Grounds in Pacific Grove, California, to hash out proposed standards for how the new tools of recombinant DNA should and should not be used.[8] Setting these types of standards was critical as the applications of GMO technologies increased.

In 1982, the FDA approved the sale of insulin made from genetically modified bacteria to diabetic consumers. Two years later, the U.S. Department of Agriculture approved the sale of Calgene's Flavr Savr tomatoes, genetically modified to ripen more slowly and last longer on the shelves. In the ensuing years, the U.S. market for genetically modified seeds grew exponentially, driven by powerful multinational corporations like Bayer, Dow, DuPont, Monsanto, and Syngenta. By 2000, most of the corn, soybeans, and cotton grown in the United States was genetically modified.

Genetically modified crop adoption levels kept growing because farmers believed that planting GM seeds increased yields and profits and reduced the need for pesticides. The U.S. public also largely came along. A series of Pew polls between 1999 and 2003 found that the majority of Americans knew little about the genetic modification of crops, about a third were concerned by it, and about a fifth thought it was safe.[9] By 2016, most Americans remained largely uninformed about GM crops, the same third of Americans polled were concerned about GMOs, and 58 percent believed that GM crops were either equivalent to or better for people's health than non-GM crops.[10] A 2016 business report found that over two-thirds of all U.S. grocery shoppers were not willing to pay any extra for nongenetically modified foods and projected that "with the continuous increase in the pricing of non-GM food products, consumers will switch to GM food products."[11]

China, too, has steadily adopted genetically modified crops. With only 9 percent of the world's arable land but 20 percent of its population, the country has always been food-insecure. For the Chinese government, genetic modification has long-seemed an appealing way to increase the crop yields of smaller and more marginal farms, grow the cotton needed for the country's massive textile operations, and feed China's people and massive herds of livestock. Today, around 60 percent of global soybean exports go to China, nearly all of it genetically engineered.

Recognizing the critical importance of GMO and other agricultural technologies to their future, the Chinese government classified "enhanced agriculture" as a strategic emerging industry in the government's most recent Five-Year Plan.[12] China "must boldly research and innovate, dominate the high points of GMO techniques," Chinese president Xi Jinping said in 2013, and "cannot let foreign companies dominate the GMO market."[13] This government push was the animating force behind the $43 billion 2017 acquisition of the Swiss multinational corporation Syngenta, one of the world's leading agricultural biotechnology behemoths, by the Chinese State-Owned Enterprise ChemChina.[14]

Even though Chinese public concern about GM has increased in conjunction with the country's repeated food-safety scandals,[15] polls show that a significant majority of Chinese consumers, like their American counterparts but unlike the Europeans, generally accept genetically modified foods.[16]

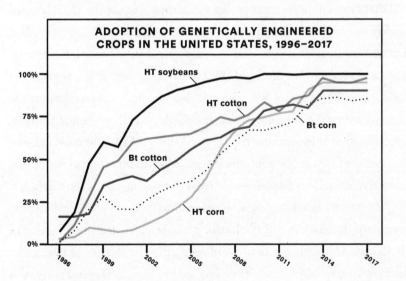

ADOPTION OF GENETICALLY ENGINEERED
CROPS IN THE UNITED STATES, 1996–2017

Source: "Recent Trends in GE Adoption," USDA Economic Research Service, last updated July 12, 2017, https://www.ers.usda.gov/data-products/adoption-of-genetically-engineered-crops-in-the-us/recent-trends-in-ge-adoption.aspx.

The growing GM adoption rate in the United States and China isn't just based on consumer ignorance. It is also based on science. As the prevalence of GM crops has grown, researchers have acquired an ever-expanding data set they have used to evaluate safety risks. Study after study over decades has repeatedly shown that genetically modified crops are as safe as conventional ones. The 2012 statement of the American Association for the Advancement of Science Board of Directors found that "crop improvement by the modern molecular techniques of biotechnology is safe."[17] The 2013 literature review in the *Journal of Agricultural and Food Chemistry* found "overwhelming evidence...[that GM crops are] less disruptive of crop composition compared with traditional breeding, which itself has a tremendous history of safety."[18] The European Union Europe 2020 report[19] as well as studies by the World Health Organization,[20] the American Medical Association,[21] the U.S. National Academy of Sciences,[22] the British Royal Society,[23] and other of the most respected organizations in the world have all reached the same conclusion.

If that's not enough evidence that GM crops are safe, the U.S. National Academies of Science, Engineering, and Medicine in 2016 published their comprehensive review of the science of GMOs. This massive meta-analysis reviewed all credible GMO studies to date from around the world, consulted hundreds of the top experts, and received comments from more than seven hundred concerned individuals and organizations. Based on all of these inputs from around the world in the most systematic review of GM crops ever, the National Academies report found "no conclusive evidence of cause-and-effect relationships between GE [a.k.a. GM] crops and environmental problems" and "no evidence of...[any] increase or decrease in specific health problems after the introduction of GE foods." Although the report did recognize the danger of herbicide-resistant weeds and some other challenges, the overall message was clear: genetically modified crops are as safe for human consumption as non-GM crops.[24]

An Italian study released in February 2018 went a step further.

Reviewing more than twenty years of data from multiple studies around the world, the authors concluded that genetic modification actually increased yields and reduced carcinogenic toxins in corn.[25] Genetically modified corn wasn't just safe for human consumption, it was according to this study even healthier for us than non-GMO corn.

That GM crops are safe to eat, however, does not mean there are not legitimate issues of concern. If we over-rely on certain crops or allow corporations to gain monopoly powers over our food supply or inadvertently create super-resistant pests, we could have real problems. GM is a tool with a huge upside and a potential downside that requires thoughtful regulation.

Despite all of these studies showing GMOs are safe, however, the scientific story of GMOs has been increasingly overcome by overblown fears in many parts of the Western world. In 1990, American environmental activists published a report titled *Biotechnology's Bitter Harvest* condemning the growing spread of genetically engineered plants and calling on the U.S. government to cut off its support for this technology.[26] Since then, anti-GMO activists, many inspired by some combination of a distrust of new technologies, U.S. global corporations, and market capitalism more generally, as well as by the romanticizing of small farmers and fears their food supplies will be contaminated, have increasingly raised the alarm over what they started calling *Frankenfoods*.

Anti-GMO organizations have launched massive disinformation campaigns designed to counter and muffle the voices of the scientific community. Many people are susceptible to this type of disinformation because most of us instinctively resort to a "naturalistic fallacy" that nature is natural, even though our ancestors have been massively modifying much of it for thousands of years. That large and vilified multinational corporations like Monsanto stand to make significant profits from this burgeoning industry also didn't help.[27]

This activism struck a particular chord in Europe. By 2016, 84 percent of Europeans polled had heard of genetically modified foods and a full 70 percent concluded that GM foods were "fundamentally unnatural."

Sixty-one percent believed that the development of GM crops should be discouraged and 59 percent that GM foods were unsafe.[28]

As these voices of public dissent became louder, European regulators listened. Even though Belgian scientists were pioneers of modern plant genetic engineering in the 1980s, a decade later European regulators became the first to require that GM foods be labeled. Labeling may superficially seem a good idea but is inherently dishonest because so many of the enzymes in foods we eat and the crops that feed our livestock are genetically modified; accepting full labeling of GMOs would require labeling much of what we eat.

In spite of the many potential benefits of GM crops, anti-GMO organizations like Greenpeace, Earth Liberation Front, and others have repeatedly and forcibly disrupted GMO research and destroyed GM crop trials in research institutes around the world.[29] In 2011, Greenpeace admitted its members had broken into an Australian research facility to destroy experimental, genetically modified wheat crops. In 2013, activists partnering with Greenpeace uprooted research paddies of the vitamin-enhanced Golden Rice at the renowned International Rice Research Institute in the Philippines.[30]

Source: Becker1999/flickr, public domain via Creative Common 2.0.

As pressure on the European Union grew, European leaders recognized that the clash of public opinion on the one hand and science and economic competitiveness on the other was creating an untenable situation. In an effort to escape the public-opinion straightjacket, EU environment ministers in 2013 agreed that individual EU countries could each decide for themselves whether to restrict GM crops for any reason. Responding more to public opinion than science, seventeen European countries banned the cultivation of GM crops by 2015.

As former anti-GMO campaigner turned GMO-advocate Mark Lynas wrote at the time:

> In effect, the Continent is shutting up shop for an entire field of human scientific and technological endeavor. This is analogous to America's declaring an automobile boycott in 1910, or Europe's prohibiting the printing press in the 15th century.[31]

A European Academies' Science Advisory Council declared that the EU was "falling behind international competitors in agricultural innovation," which had "implications for EU goals for science and innovation, and for the environment as well as for agriculture."[32] By then, it was too late; the European anti-GM train had already left the station.

In Berlin in 2015, I met with a number of the German regulators involved with enforcing the country's restrictions on GM crops. Every one of them told me they thought the restrictions were unscientific, counterproductive, and asinine. All of them said they were required to restrict GMOs against their better judgement, because public opinion had forced the hand of Germany and Europe's political leaders.[33] A few described how Greenpeace, the leading anti-GMO campaigner, had allegedly lobbied against the labeling of genetically modified enzymes used in most cheeses, breads, wines, and beers as GMOs out of an apparent fear that if people understood how reliant we are on GMOs, the anti-GMO campaign would collapse.[34] And even though new gene-editing techniques like CRISPR are making it possible to turn on or off already existing genes within crops to alter traits, and are therefore not technically GMOs, Greenpeace and the anti-GMO forces are beginning to rally in opposition.[35]

In 2016, 109 Nobel laureates issued an open letter calling on Greenpeace to end its anti-GMO campaign. "We urge Greenpeace and its supporters," the letter said, to:

> re-examine the experience of farmers and consumers world-wide with crops and foods improved through biotechnology, recognize the findings of authoritative scientific bodies and regulatory agencies, and abandon their campaign against "GMOs" in general and Golden Rice in particular. Scientific and regulatory agencies around the world have repeatedly and consistently found crops and foods improved through biotechnology to be as safe as, if not safer than those derived from any other method of production. There has never been a single confirmed case of a negative health outcome for humans or animals from their consumption. Their environmental impacts have been shown repeatedly to be less damaging to the environment, and a boon to global biodiversity.[36]

Despite the impassioned pleas of the expert community, the impact of the anti-GMO campaign remains high, particularly in Europe. These efforts scored a major victory in July 2018, when Europe's highest court, the Court of Justice of the European Union, ruled that gene-edited crops altered with new gene-editing techniques like CRISPR would be subject to the same strict regulation as GM crops, even though no outside genetic material was being introduced.[37] "This will have a chilling effect on research, in the same way that GMO legislation has had a chilling effect for 15 years now," Umeå University plant physiologist Stefan Jansson told Nature.[38]

Europe's GMO bans hurt the continent's economic competitiveness but don't risk the lives of any European citizens. The same could not be said for the impact of these policies on the developing world. European restrictions on certain agricultural imports from countries with GM crops forced many African and Asian governments depending on those export markets for their economic well-being to themselves restrict planting genetically modified crops. This denied poor countries the opportunity to use biotechnology to generate bigger yields of virus-resistant crops, reduce use of fertilizer and dangerous pesticides, add life-saving nutrients to their diets, and help minimize the impact of droughts that can cause thousands of deaths, hardly a benign outcome.[39]

The Nobel laureates reserved their harshest criticism of Greenpeace for its opposition to Golden Rice. "Greenpeace has spearheaded opposition to Golden Rice," they wrote:

> which has the potential to reduce or eliminate much of the death and disease caused by a vitamin A deficiency (VAD), which has the greatest impact on the poorest people in Africa and Southeast Asia. The World Health Organization estimates that 250 million people suffer from VAD, including 40 percent of the children under five in the developing world. Based on UNICEF statistics, a total of one to two million preventable deaths occur annually as a result of VAD, because it compromises the immune system, putting babies and children at great risk. VAD itself is the leading cause of childhood blindness globally affecting 250,000–500,000 children each year. Half die within 12 months of losing their eyesight. WE CALL UPON GREENPEACE to cease and desist in its campaign against Golden Rice specifically, and crops and foods improved through biotechnology in general... How many poor people in the world must die before we consider this a "crime against humanity"?[40]

Although Africa's restrictions on GM crops have recently begun to loosen, there is little doubt that European and activist NGO scare-mongering has delayed the adoption of GM crops in Africa and elsewhere.[41] The application of genetic modification and gene editing to crops has the potential to significantly improve the resilience and sustainability of our food supply and make living on our warming planet with its growing human population more feasible and survivable.[42] While the arguments of the anti-GMO campaigners are not entirely without merit and should be considered, there is little doubt that the scare-mongering anti-GMO campaign is doing far more harm than good.

The GM crops experience shows how heated the debates over altering what different people perceive differently as "nature" can become.

The GMO debate also gives us a good indication of what might be heading our way when the organisms being genetically modified are not soybeans and corn but us. And we've already seen how sensitive and volatile ideological divisions can become when human reproduction is at stake.

X

Humans have been performing abortions for most of our recorded history. Early records of abortions go back to 1550 BC in Egypt. Abortion was commonplace and widely accepted in ancient Greece and Rome. Plato and Aristotle both explicitly endorse in their writings the right of women to receive abortions. Chinese archives from as early as 500 BC describe mercury being used to end pregnancies.

In 1803, the British parliament made abortion after five months of pregnancy—when it was then believed the soul entered the fetus— punishable by execution. The penalty was later reduced to life imprisonment for whoever performed the abortion. Although abortion was common in Colonial America, restrictions in the United States started in the 1850s, forcing American women who sought to terminate their pregnancies to have even less sanitary and more dangerous abortions than the already low medical standards of the day and leading to thousands of unnecessary deaths. Although the UK Abortion Act of 1967 and the landmark 1973 U.S. Supreme Court decision in the case of

Roe v. Wade established a women's right to an abortion in both countries, the abortion issue was far from settled.

In the years following *Roe v. Wade*, the Catholic Church and evangelical Christians rallied against abortion with increasing intensity, eventually gaining the political support of America's Republican Party. Although the *Roe* decision held and abortion rights remained protected in more liberal states like California and New York, pressures to restrict abortion intensified in more conservative U.S. states. In the forty years after *Roe*, nearly 300 violent attacks, including arson and bombings, were carried out against abortion clinics, and abortion providers were even murdered across the United States.[43] The National Abortion Federation additionally reported over 176,000 incidents of abortion clinics being picketed over the past four decades, 1,500 acts of vandalism, and 400 death threats.[44] As a result of these types of pressures and the 1977 Hyde Amendment that significantly restricted the use of federal funds for abortion, today 84 percent of U.S. counties provide no abortion services, and only seventeen U.S. states fund abortions on terms similar to other health services.[45]

In China, the Communist party quickly banned most abortions after coming to power in 1949 to spur population growth. But after taking over in 1978 following Mao's death and a power struggle, Deng Xiaoping believed his economic reform plans could only work if China's population growth could be slowed. His government launched the one-child policy in 1979, limiting most Han Chinese families, particularly those in urban areas, to a single child. In support of this policy, China legalized abortion in 1988.

Over the following nearly four decades the one-child policy was in force, most families that had more than one child were heavily fined and a large number of women were forced to have abortions and/or be sterilized without their consent.[46] The one-child policy eased any stigma surrounding abortion and encouraged many parents, sometimes inadvertently and sometimes directly, to abort female embryos, place female babies for adoption, and abandon children born with disabilities.

Although the one-child policy was finally eased in 2015, the policy subtracted around four hundred million people from China's otherwise predicted population growth.[47]

But even with government pressure relaxed, the Chinese public remained on average considerably more comfortable with abortion than their American counterparts. Unlike Western religions that traditionally ascribe humanity as a divine creation, the Chinese tend to see themselves as descendants of their own parents, a foundation of the ancient rite of ancestor worship. These ideas also underpin the Confucian concept of filial piety, a major foundation of Chinese culture, which highly values producing healthy and capable offspring to carry on the family legacy. Today, abortion remains widely accepted in China.

Around the world, these types of different cultural traditions and legal norms have inspired different levels of acceptance and comfort with abortion. The chart below outlines the differences between various religions' position on abortion based on Pew's 2014 Religious Landscape Study:[48]

MAJOR RELIGIOUS GROUPS' POSITIONS ON ABORTION

Opposes abortion rights, with few or no exceptions	Supports abortion rights, with some limits	Supports abortion rights, with few or no limits	No clear position
• African Methodist Episcopal Church	• Episcopal Church	• Conservative Judaism	• Islam
• Assemblies of God	• Evangelical Lutheran Church in America	• Presbyterian Church (U.S.A.)	• Buddhism
• Roman Catholic Church	• United Methodist Church	• Reform Judaism	• National Baptist Convention
• Church of Jesus Christ Latter-day Saints		• Unitarian Universalist	• Orthodox Judaism
• Hinduism		• United Church of Christ	
• Lutheran Church-Missouri Synod			
• Southern Baptist Convention			

Source: David Masci, "Where Major Religious Groups Stand on Abortion," Pew Research Center, June 21, 2016, http://www.pewresearch.org/fact-tank/2016/06/21/where-major-religious-groups-stand-on-abortion/.

These types of different communal views on abortion show up in public opinion polling from around the world. In 2017, for example, 58 percent of Americans polled believed abortion should be legal or mostly legal while 40 percent believed it should be illegal or mostly illegal. Not surprisingly, Jewish, Buddhist, and unaffiliated Americans supported abortion rights the most, while Catholics, Evangelicals, Mormons, and Jehovah's Witnesses opposed most strongly.[49] A 2015 BuzzFeed poll of twenty-three countries found, not surprisingly, that liberal European countries were most permissive of abortions while more Christian and traditional Latin American, Asian, and African countries less so.[50]

These differences in belief show up in the wide range of laws regulating abortion in different jurisdictions around the world. According to Pew's 2015 analysis of global abortion laws, 26 percent of the 196 countries surveyed allow abortions only to save the life of the mother, and an additional 42 percent place significant restrictions on abortion. Most of these restricting countries are ones where religious institutions play a leading role in society. All of the six countries banning abortion under any circumstances are countries where the church dominates.[51] These different legal structures explain why women living in Louisiana, one of the U.S. states with the toughest restrictions on abortion,[52] travel out of state when they need abortions or why women from Andorra, where abortions are banned, cross the border to France when seeking theirs.

The disparities between how different communities approach and regulate abortion is not just a precursor to the genetically modified humans debate but a central component of it. Because IVF and embryo selection, the gateway procedures for heritable human genetic engineering, nearly always entail the destruction or at least permanent freezing of unimplanted embryos, the politics of abortion is already morphing into the politics of assisted reproduction, embryo screening, and human gene editing. America's Hyde Amendment is a perfect example of this. Inspired by the abortion debate, the amendment also created significant restrictions on human genetics research in the United States.[53]

If people have gone ballistic over the environmental, GMO, and abortion debates, if they have manned the barricades and destroyed research centers over genetically modified crops or attacked abortion clinics and murdered doctors, imagine what they might do when the same diversity of individual, cultural, societal, and governmental views inspires different national approaches to the emerging science of genetically modified people.

X

In 2016, a French activist group called Alliance VITA launched a campaign to "Stop GM Babies" with the petition below.[54]

PETITION

CRISPR-Cas9:
Yes to therapeutic progress, no to transgenic embryos!
Men deserve to be cared, and not genetically programmed.

For the past few months, the use of the CRISPR-Cas9 technology has been sharply increased: this technique allows to modify directly the DNA (genome) of any vegetable, animal, or human cell.

This technology, is promising to treating various genetic diseases in children and adults. **But when applied to human embryos, it can produce from scratch, genetically modified humans**: "GM babies."

A worldwide ethical regulatory framework must be implemented.

GM babies? NO!

By signing this petition, I ask my country to urgently undertake and secure an international moratorium—meaning an immediate halt—on genetic modification of human embryos, **in particular using the CRISPR-Cas9 technology**.

Source: Alliance VITA, "Stop GM Babies: A National Campaign to Inform and Alert about CRISPR-Cas9 Technique," May 24, 2016, https://www.alliancevita.org/en /2016/05/stop-gm-babies-a-national-campaign-to-inform-and-alert-about-crispr-cas9 -technique/.

Signed by more than ten thousand people, this petition was not a wild success but was a harbinger of things to come. The future debate over genetically modified humans could be, at least in some ways, less contentious than the abortion struggle if the benefits of human genetic engineering can be widely experienced by the public more quickly than the fear-mongering can take hold. IVF is a good example of this.

In 1978 just after the birth of Louise Brown, 28 percent of Americans polled said IVF was morally wrong. Relatively quickly, however, people started to witness the benefits of IVF for helping women have babies. This evident success is why we are more likely to see pro-life campaigners picketing abortion clinics than fertility clinics, even if far more early-stage embryos are terminated in the latter. When asked in 2013, only 12 percent of Americans said IVF was morally wrong,[55] far less than the 40 percent opposing abortion. On the other hand, public attitudes toward preimplantation genetic testing and genetically altering embryos have not yet normalized like they have for IVF.

Seventy-four percent of American surveyed in a 2002 Johns Hopkins University poll expressed support for using preimplantation embryo screening to avoid serious disease. But only 28 percent were comfortable using it to select the gender of a child, and 20 percent approved of using it to select for desirable non-disease-related traits like intelligence.[56]

Gene editing preimplanted embryos is, of course, a more significant intervention than selecting which embryo to implant in the mother. When asked by STAT and Harvard's T. H. Chan School of Public Health in 2016 whether they thought gene editing should be used to improve the intelligence or physical characteristics on an unborn child, only 17 percent of Americans agreed.[57] A Pew poll two years later similarly found that only 19 percent of Americans would be willing to change their baby's genes to make the baby more intelligent.[58] But when asked if they would be willing to gene edit their own future children to significantly reduce their baby's risk of serious disease, nearly half of the Americans polled said yes in 2016, and that number increased to 72 percent in the 2018 poll.[59]

These polls not only show that Americans are more comfortable with interventions having a clear medical purpose than with everything else but also that parents will, conceptually at least, aggressively protect their future children from risk, including by altering their children's genetics. The polls also indicate that popular American comfort levels with these types of interventions are growing. Like with the abortion polling, religious Americans are far more restrictive in their attitudes toward gene editing embryos than are the nonreligious.

The United Kingdom has seen perhaps the greatest shift toward increased acceptance of more aggressive genetic interventions. A third of Britons polled in 2001 felt that genetics research writ large was unethical, and tampering with nature.[60] By 2017, however, 83 percent of UK residents surveyed by the Royal Society supported gene editing to cure serious disease when the genetic changes would not be passed to future generations. Seventy-six percent supported editing human genes to correct genetic disorders even when the genetic changes would be passed to future generations. An astounding 40 percent supported using genetic engineering to enhance human abilities like intelligence.[61]

The British population has experienced the highest level of public education on genetic technologies in the world, particularly in the context of the national conversation on mitochondrial transfer. This, plus the United Kingdom's pro-science and open-minded national orientation, have translated into over three-quarters of the population supporting the genetic alteration of embryos in ways that would be passed to future generations forever, and nearly half the population expressing a willingness to genetically enhance their own future children.

Although China was far behind the West in assisted reproduction technologies only a decade ago, the country is showing the biggest global swing toward widespread acceptance of assisted reproduction. Living in a society where religious faith has been suppressed for decades, screening and aborting embryos is seen as less of a religious issue in China than in most other countries. The ancient Chinese concepts of *taijiao*, or fetal education, and *yousheng*, healthy birth, emphasize the importance of birthing optimally healthy babies.[62] Along with the significant Chinese cultural stigma surrounding and lack of institutional support for people with disabilities, these concepts have paved the way for a greater societal acceptance of embryo screening in China than in most other countries.

In 2004, only four clinics in China had a license to perform PGT. By 2016, the number had risen to forty. Forty may not seem like a big number for a country as vast as China, but many of these clinics are now operating on a colossal scale far beyond clinics anywhere else. One single clinic in Changsha, a city near Beijing, reported 41,000 IVF procedures in 2016, a quarter of the total number carried out in the entire United States and more than in the entire United Kingdom that same year.

Across China, PGT is growing by an average of 60 to 70 percent annually. Costing a third of the U.S. price, the total number of these procedures in China is already greater than that in the United States and climbing.[63] Chinese clinics are advertising their ability to use PGT to eliminate the risk of a growing number of genetic diseases. When asked in a 2017 online survey, the Chinese people surveyed on average "mildly agreed" they would be comfortable genetically altering their future children.[64]

These societal difference between the United States, the United Kingdom, and China may ultimately end up being less significant than the generational differences where younger people appear more comfortable with genetic engineering than their elders.[65]

As in the abortion context, religious differences are also already pushing communities in different directions on these issues. Most but not all religions accept that gene therapies can and even should be used to treat and cure diseases when genetic changes are not passed to future generations. But the holy water gets murkier from there.

The Catholic Church strongly opposes the use of PGT in conjunction with any type of embryo selection.[66] In his 1995 *Evangelum Vitae*, Pope John Paul II wrote that prenatal diagnostic techniques incorporate a "eugenic intention" that "accepts selective abortion in order to prevent the birth of children affected by various types of anomalies." This, he wrote, "is shameful and utterly reprehensible, since it presumes to measure the value of a human life only within the parameters of 'normality' and physical well-being, thus opening the way to legitimizing infanticide and euthanasia."[67] In 2013, the Catholic Church opposed the draft British law authorizing clinical trials of mitochondrial transfer, citing a 1987 Vatican instruction that "medical research must refrain from operations on live embryos, unless there is a moral certainty of not causing harm to the life or integrity of the unborn child."[68]

The National Council of Churches of Christ, a more progressive protestant group, expressed in its 2016 report a greater appreciation for the role heritable genetic manipulation might play. "Effective germ line therapy could," the report said, "offer tremendous potential for eliminating genetic disease, but it would raise difficult distinctions about 'normal' human conditions that could support discrimination against people with disabilities."[69]

But that's not as far as Christian liberalism extends. At a 2017 conference on the future of AI, I attended at Ditchley Manor in Oxfordshire, England, I met a fascinating Presbyterian minister, Christopher Benek, who describes himself as a "techno-theologian" and "Christian Transhumanist." Humans are "co-creators with Christ," he argued, and stewards of technologies like AI and gene editing that can be used to improve upon humanity. In his writing, he calls for an "ethical and helpful use of emerging technologies to enhance humans."[70]

Chris's views are certainly at the fringes of Christian theology, but they are probably a little closer to the center of gravity of Jewish thought. The traditional Jewish concept of *Tikkun Olam*, or *repair the world*, suggests that the world is broken, and it is the responsibility of each Jew, as god's agent, to help fix it. "All that was created during the six days of creation," the Torah says in Genesis 11:6, "requires improvement... the wheat needs to be ground, and even a person needs improvement." Completing god's creation, in this sense, is seen not as critique of the divine but as embracing man's divine purpose. Judaism doesn't have the type of hierarchical command structure as Christianity, but even the more conservative Orthodox Jewish communities were, as the Tay-Sachs example shows, the earliest adopters of advanced genetic technologies.

In a 2015 article in a Jewish medical journal, Rabbi Moshe Tendler of Yeshiva University and bioethicist John Loike make the case for why Jewish law supports Jewish women participating in clinical trials of mitochondrial replacement therapy (MRT), even though this treatment could ultimately result in the birth of genetically altered babies. Jewish law, they wrote, "permits women to engage in new biotechnologies such as MRT to have healthy offspring...[and] volunteering for such trials is not only a *chesed* [an act of loving-kindness] but engenders social responsibility so that Jews are contributing to the overall health of our society."[71]

Being open to some genetic alterations does not mean, of course, that transhumanist Christianity, mainstream Judaism, progressive Buddhism, and other communities of faith are signing up for the slippery slope to unlimited transhumanism. But the diversity of beliefs on human genetic enhancement is already significant within and between religious and intellectual traditions and is more likely than not to grow as genetic technologies come of age. As this happens, science in general and transhumanist aspirations in particular will increasingly take on some of the trappings of a religion.[72]

Even in this human genetic revolution, these types of significant differences within and between individuals, communities, and governments about heritable genetic engineering are beginning to play out in a great diversity of national laws.

There is no U.S. federal legislation covering IVF in the United States, and America also has among the most accommodating regulatory structures for PGT. U.S. federal law bans funding for research in which embryos are destroyed, but the ways in which genetic screens are used during PGT is almost entirely left to the discretion of doctors themselves, guided by their medical professional organizations. As a result, an estimated 9 percent of all PGT carried out in the United

States is for gender selection, which is illegal in countries like China, India, Canada, and the United Kingdom. A small number of U.S. clinics even allow prospective parents to affirmatively select embryos to ensure children *will* have genetic disorders like dwarfism or deafness, unthinkable in other parts of the world.[73]

No common legislation governs assisted reproduction in Europe, resulting in a patchwork of different laws and regulations. Most European states stipulate that PGT can only be used to select against serious and incurable diseases. Some countries, like Italy and Germany,[74*] are more restrictive, while others, like France and the United Kingdom, have established broader regulatory bodies, the *Agence de la Biomédecine* in France and the United Kingdom's Human Fertilization and Embryology Authority, that tend to be more permissive when reviewing PGT applications.[75] Because a 2008 European Commission directive makes all Europeans free to travel without penalty to other EU countries for treatments, including PGT, that may be restricted at home, the impact of various national restrictions is muted.

In countries with strong religious leanings, including Chile, the Ivory Coast, the Philippines, Algeria, Ireland, and Austria, PGT is banned. If someone wants to get PGT in Chile or select a boy in China, they can find an underground clinic, travel to another country, or simply not have it done. A recent survey showed that 5 percent of those receiving assisted reproduction care in European clinics and 4 percent in the United States were people traveling from abroad to circumvent restrictions and other barriers in their home countries.[76]

Among the more advanced countries, Australia, Belgium, Brazil, Canada, France, Germany, and the Netherlands prohibit gene editing

---

* Until relatively recently, Germany had among the most restrictive laws in the world protecting the rights of human embryos, stemming largely from the country's principled reckoning with its murderous Nazi past. Germany's 1990 Embryo Protection Act forbade the fertilization of human eggs for research and embryos from being donated. In response, German prospective parents often traveled to Belgium to have PGT carried out. As public notions of parenting changed, however, the German government voted in 2011 in favor of allowing PGT under specific circumstances.

embryos in ways that would be passed to future generations, most imposing criminal penalties against violators. Rather than banning germline manipulations up front, a second set of countries—including France, Israel, Japan, and the Netherlands—make it illegal to initiate a human pregnancy with a genetically modified embryo. A third set of countries, including the United Kingdom, carves out very specific exceptions to a ban and establishes regulatory structures for making case-by-case decisions about what types of heritable changes can be allowed.

The United States doesn't have a ban on heritable genetic alteration of embryos per se but has other regulatory structures that make this type of activity all but impossible. China, on the other hand, has decent laws restricting heritable human genetic manipulation but a weak and often inconsistent oversight culture, a Wild West mentality among some researchers, and a very large, unlicensed "gray market" of assisted reproduction clinics.[77] That's why it was no surprise the highly controversial alleged first gene editing of the human embryos apparently taken to tern in November 2018 happened in a Chinese laboratory with no oversight from the university or government. Other countries have no significant laws on human genetic alteration; some among these are positioning themselves as future destinations for unrestricted human reproductive tourism.[78]

The maps on the following pages gives some indication of the diverse regulatory environment for genetic technologies around the world:

HUMAN GERMLINE GENETIC MODIFICATION

HUMAN REPRODUCTIVE CLONING

HUMAN SOMATIC GENE THERAPY

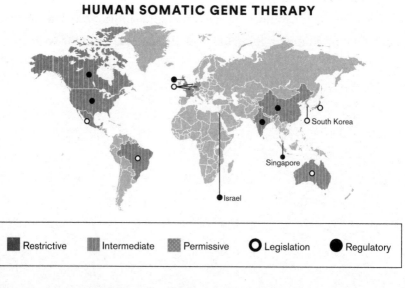

PREIMPLANTATION GENETIC DIAGNOSIS

HUMAN RESEARCH CLONING

HUMAN EMBRYONIC STEM CELL RESEARCH

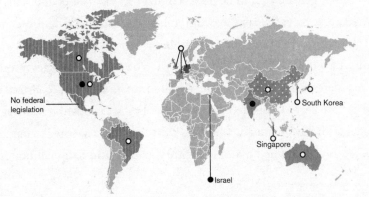

Source: R. Isasi, E. Kleiderman, and B. M. Knoppers, "Editing Policy to Fit the
Genome?" *Science* 351 (2016): 337–339.

Because genetic data will be so critical to both the personalized health care of individuals and the deciphering of human genetics more generally and also so susceptible to misuse by others, societies will have good reason to safeguard the genetic information of its citizens. Here again, national variation is big and growing. Different countries regulate and protect the privacy of genetic and other data very differently.

America's 2008 Genetic Information Nondiscrimination Act (GINA), Canada's Genetic Non-Discrimination act, the United Kingdom's 2010 Equalities Act, and Australia's Genetic Privacy and Non-Discrimination Bill prohibit discrimination based on genetic information in health insurance and employment but provide far less sweeping consumer genetic data protections than in the European Union.*

The EU has gone the farthest in protecting individual privacy rights. In May 2018, the EU's revolutionary General Data Protection Regulation, GDPR, came into effect. By far the most aggressive data-protection law in the world, the GDPR enshrined the privacy rights of EU citizens to control access to data collected on them anywhere in the world and placed strong new obligations on companies to safeguard that data. Although the international debate over GDPR has largely centered on the pressure it places on U.S. tech giants like Google and Facebook, the implications for genetic privacy are equally significant. Under the GDPR, each subject must give explicit consent for his or her genetic data to be included in a specific data pool or study.[79]

China, too, recently passed a comprehensive national data protection law with big implications for genetic information. The 2017 Network Security Law of the People's Republic of China imposed privacy protection mandates on companies and required an individual's consent for any data transferred outside of China. The law also bans the transfer of personal information outside of China if the transfer is deemed to damage Chinese vaguely defined national security, public, and national interests.

---

* Interestingly, GINA does not apply to people in the U.S. military or to other forms of insurance, such as life insurance. Genetic nondiscrimination will become particularly important because universal sequencing will show that everyone has a preexisting condition of some type or an increased risk for multiple disorders relative to the general population.

This requirement applies to what the law calls "sensitive" personal information, a category including genetic data.[80]

Although Europe and China's privacy laws may at first seem similar, they are anything but. In Europe, the GDPR is being deployed to protect individual citizens from their data being used without their consent by anyone, including the government. In China, the new privacy law is ensuring a government monopoly over genetic and other data collected from and about people inside China.

While the EU is working to prevent genetic data from being used to create a police state, China's Ministry of Public Security is amassing the world's largest DNA database as, among other things, an investment in social control. Chinese citizens in the country's predominantly Muslim Xinjiang region and other ethnic minorities, migrant workers, potential dissidents, college students, and activists are being required to provide samples to be entered into the national database.[81]

Privacy protection may seem like a justifiable human-rights issue in the EU and an Orwellian control issue in China, but the reality, as always, is more complex.

For most people, the idea of a government or company tracking every aspect of our lives is frightening. From an individual rights perspective, privacy is essential. From a big-data-analytics perspective, however, it has the potential to be a barrier to pulling together the vast data sets from which actionable insights can be drawn.

Because understanding what individual and groups of genes do is the ultimate big-data problem, the larger and higher quality the data sets, the more possible it will become to uncover genetic patterns underpinning more complex diseases and traits. Whoever gets the biggest and best data sets will be best poised to lead the genetic revolution with all of the wealth, prestige, power, and influence it will bring.

It may end up being the case that China's massive investment in AI, efforts to promote national leadership in life sciences and biotechnology, and ability bring together massive data sets of sequenced people paired with full access to their electronic medical records will put China in the pole position in global efforts to decode the human genome, transform health care, and lead the global genetics revolution. It could alternately be the case that greater privacy protection in the United States and Europe will lead to stronger and more coherent societies, higher standards of genetic and other data collection, and more quality discoveries. Whichever society makes the right bet will be poised to lead the future of innovation. But we should not delude ourselves about what's at stake and the societal costs of making the wrong bet.

X

The environmental, GM crops, abortion, and genetically engineered humans debates show how our different communities, with their diverse histories, cultures, economic pressures, and political structures, respond very differently to new technologies. These dissimilarities then lead to a wide variety of legal and regulatory environments around the world.

The good news is that the wide disparity of different approaches is creating a "laboratory of nations," each finding its own path as it interacts and competes with other states. In this context, the most promising, and sometimes even the most aggressive, research on and applications of any new technologies will find a home, driving innovation forward.

The potentially bad news is that the diversity of communal beliefs and national models could also move humanity toward the lowest common denominator approach to human genetic engineering. If this should happen, the most aggressive countries will set the bar for everyone else, who must keep up if they believe their future well-being, competitive advantage, and prosperity are at stake. Our diversity of approaches, like all forms of diversity, increases the odds of both positive and negative outcomes.

Diversity, however, is a biological precondition but does not alone drive evolution. Evolution also needs that other essential ingredient: competition.

If our different orientations about human genetic engineering establish a wide range of options, competition between us will propel our species into the genetic age.

Chapter 10

The Arms Race of the Human Race

When given the opportunity to gain some type of advantage over others, even at a considerable risk, some subset of us takes it. We've evolved this way.

From the moment life emerged, our ancestors entered a never-ending arms race with each other and other species for advantage and survival. We fought our way to the top of the food chain and avoided being ground up as burgers for some other species because we've (so far) won that competition.

In our nomadic, hunter-gatherer days, human groups in many places competed relentlessly with each other, often stealing each other's resources. When the advent of agriculture, writing, and other technologies made it possible to organize ourselves into larger communities, we wasted no time translating each little technological advantage into individual and collective opportunities to rob, subjugate, and oppress each other.

The Mongols leveraged the stirrup to give their horsemen additional oomph to conquer much of the known world. The European colonial powers used their advanced ships and weapons to dominate and exploit big portions of the globe. The initial technological advantage the Germans and Japanese enjoyed in the Second World War was overcome by the even more powerful competitive advantage of the American, British, and

European émigré scientists who developed the radar, advanced cryptography, radio navigation, and nuclear weapons that helped win the war. In these cases, and so many more, competition fueled technological development even when the technologies being developed had both great upsides and dangerous potential downsides.

And although utopians for many centuries have imagined a world where we all fully embrace our Buddha-nature and escape the cycle of incessant competition with each other, that day has not yet arrived. Even though levels of large-scale conflict have been decreasing around the world for decades,[1] it would be dangerous folly to believe that the genetics revolution with its great upsides and dangerous potential downsides will play out in the harmonious, noncompetitive world we can imagine rather than the highly competitive one we have known. Instead, the magnitude and consequences of this competition will grow as the technology advances.

In the early years of this expanding revolution, particularly when our understanding of what genes do, how the body works, and what manipulations are most beneficial is in its infancy, an artificial distinction between therapy and enhancement will be maintained. We will be able to maintain the fiction that the genetics revolution is primarily and ultimately about enhancing health care and treating disease. In the early years, doctors working to prevent or treating a certain disorder or disease may want to overshoot the mark of normal by giving an embryo (or even a grown person) extra advantages. Because what is considered a therapy to one person might be considered an enhancement to another, our concept of what is "normal" will prove a moving target. As therapeutic applications of powerful genetic technologies become the norm, the distinction between what constitutes a therapy and a genetic enhancement will increasingly blur.

We will ask, to give a few examples, whether there's really a fundamental difference between genetically enhancing someone's poor eyesight to be normal versus their normal eyesight to be great, or if there's really a difference between genetically enhancing a patient's cellular ability to fight a cancer or AIDS that's already developed or their ability to never get cancer or AIDS in the first place. Parents, in other words, will increasingly wonder if there's really a difference between giving their children advantages of nurture versus providing the same advantages through nature. The more enhancements like these become available and seen as beneficial, the more competition within and between communities will push our highly competitive fellow humans to want them.

The sporting world provides a particularly illuminating window into how competitive pressures can and likely will spur a genetic arms race.[2]

X

There is little doubt that genetics play a central role in achievement in sports at the highest levels, where the rules are clearly defined, the capabilities needed for success are relatively specific, and the distribution of nongenetic assets are evenly distributed. That's why there are so few five-foot players in the National Basketball Association.

As David Epstein describes in his 2013 book *The Sports Gene: Inside the Science of Extraordinary Athletic Performance*, Finland's Eero Mäntyranta was one of the most dominant Nordic skiers in history, winning seven Olympic medals, including three golds, and two world championships between 1960 and 1972. A consummate champion, Mäntyranta had a tremendous work ethic and indomitable spirit. When he and his family were genetically sequenced in the early 1990s, however, it turned out that Mäntyranta and twenty-nine of his relatives also had a very rare single mutation of the EPOR gene. This mutation made them far better able than others to produce hemoglobin, the red blood cells that transport oxygen from the lungs to body tissues, which gave them a far greater than normal endurance capacity.[3] Not every relative with this mutation was an Olympic champion. Few were. But having this mutation certainly increased their odds.

Athletic success is a complex phenomenon that cannot of course be attributed to any one gene, even one like Mäntyranta's. Lots of other factors like nutrition, parenting, drive, coaching, access, and luck play important roles. But that doesn't mean single genes can't be critically important.

Lots of people have a mutation of their *ACTN3* gene, one of the unknown larger number of genes influencing the speed at which muscles contract. Although many athletic performance-related genes have been studied, *ACTN3* is so far the only one that correlates with performance across multiple power sports. When researchers knocked out this single gene in mice, the mice lost significant levels of muscle power.[4] Having the right *ACTN3* variant won't make you a great sprinter, but multiple studies have shown that if you have two disrupted copies of *ACTN3* your chances of making it to the Olympic finals in any sprinting race are virtually nil.

If you have a mutation of the *MSTN* gene impeding the production of myostatin, on the other hand, a protein that halts the production of muscles, your muscles keep growing far beyond most everyone else's. Liam Hoekstra, a child with this myostatin mutation dubbed "the world's strongest toddler," could do challenging gymnastics tricks at five months old and pull-ups at eight months.

Genetics are also proving highly influential in elite marathon running. The spectacular success of Kenya's Kalenjin tribe, and even more of its Nandi subtribe, is a case in point. Most everybody knows that Kenyan runners have dominated distance running for decades. Between 1986 and 2003, the percentage of Kenyan male runners included in the top twenty fastest times ever in races 800 meters and longer increased from 13.3 percent to an astonishing 55.8 percent. Over the past thirty years, Kenyan-born men have won nearly half of all Olympic medals in distance running events and almost all of the world cross-country championships. But the Kenyan running success story is more precisely a Kalenjin success story.

About three fourths of Kenyan running champions are Kalenjins, a tiny group making up only 4.4 million of Kenya's 41 million people. Kalenjin runners won 84 percent of Kenya's sixty-four Olympic and eight world-championship medals between 1964 and 2012 and were responsible for twenty of Kenya's fastest twenty-five marathon times.[5] The majority of these Kalenjin runners come from the even smaller Nandi subtribe, a group of about a million people in total. While this running success reflects lots of hard work, a national running culture, and other environmental factors, the genetic foundation of Kalenjin dominance is hard to miss.

Kalenjin runners have on average longer legs, shorter torsos, thinner limbs, and less mass in relation to height than most everyone else, to which anyone who's ever watched a competitive marathon over past decades can attest. Swedish researchers testing Kalenjin children in 1990 found remarkably that over five hundred Kalenjin schoolboys could run two thousand meters faster than Sweden's fastest champion. After researchers from a Danish sports science institute trained a large group of Kalenjin boys for three months in 2000, many allegedly ran faster than a leading Danish runner.[6]

Although no single marathon gene has been or will likely be discovered, researchers have identified the important influence over two hundred genes can have in athletic performance.[7] A 2008 study that assessed the possibility of a given runner having each of the twenty-three best-understood of these athleticism-related genes concluded it would be extremely small, about 0.0005 percent of all people. If that's true, that would mean that China, with its 2018 population of 1.34 billion people, would have 6,700 people with these genetic attributes spread across all age groups. The United States, with its population of 324 million, would have 1,620.

Imagine a country—let's say China—decided to use the data it is already collecting when sequencing its newborns to determine which among them would have the greatest chance to be star athletes of one type or another. The genetic potential could be flagged to parents and sports associations, who might be encouraged to give these children special opportunities to participate in sports where they might have some type of advantage. Those children demonstrating the greatest aptitude and passion for a given sport could be invited to join special leagues, where the best players could be encouraged or required to attend special sports schools and receive rigorous additional training. The best among these genetic stars would be identified to represent the country internationally. A bit of this is already happening.

When attending women's volleyball semifinals in the 2008 Summer Olympics in Beijing, I noticed a strange contrast between the American and Chinese athletes. The Americans were a hodgepodge of various shapes and sizes. In the American model, young athletes are often self-starters or the children of obsessive parents. Some of the American players, I am sure, had levels of heart, determination, team spirit, or charisma that had brought them to the Olympics. The Chinese, on the other hand, were far more uniform in appearance. Like the Soviets decades before, the Chinese often identify potential athletes based on physical attributes at an early age and then train and cull them through special sports schools around the country. Critics may not like this sports-school model, particularly when families have been coerced to give up their children to these schools, but few would call identifying young athletes for athletic potential cheating.

The Beijing volleyball competition was not just between two teams but also between two different forms of societal organization. It may be that the American individualist system proves better able to create dominant athletes, the Chinese and Russian statist systems may do better, or some hybrid of the two will carry the day. Over time, however, this will not remain a theoretical point. Athletic success will be measured in the ongoing arms race of Olympic medals.

Bringing advanced genetic capabilities to the sports arms race won't just create new opportunities for cheaters but also challenge our conceptions of what constitutes fair play. People have long complained that the state-sponsored doping of athletes in the Soviet bloc countries, Russia, China, and elsewhere—and athletes like cyclist Lance Armstrong, who artificially stimulated his body's production of red blood cells to boost hemoglobin levels—were cheating. They were, according the rules of their sports.

At some point in the near future, athletes will be able to have their cells extracted and gene edited then reintroduced to their bodies to increase endurance and speed up muscular repair.[8] This type of "gene doping" will clearly be cheating, based on the current rules in most sports. But now that we can look under the genetic hood, is it fair for a person without an extra hemoglobin-producing mutation to compete against someone who has it?

What do we think about people like Mäntyranta, whose bodies are, on their own, doing what the cheaters are seeking to mimic? Are athletes like Lance Armstrong committing a violation by artificially boosting their hemoglobin production or just leveling the playing field by "Mäntyranta-ing" themselves?

The more genetic predispositions for relative success in certain sports can be identified, the less fair it will be for athletes without them. But because all humans are genetically different from each other, penalizing an athlete for having an advantageous genetic difference would be like penalizing a physicist for having a genetic predisposition for strong math skills or a musician for having perfect pitch.

Once we understand these genetic differences that give some athletes an inborn genetic advantage over others, one response might be to categorize athletes by their genetic differences, where people like Eero Mäntyranta could compete against each other, and people like Lance Armstrong could do the same. Kalenjin runners could likewise race against each other and we'd have a different competition for everyone else. This would be entirely absurd for many reasons, not least because physical diversity is the essence of competitive sports and because it would be impossible to determine which particular set of genetic attributes created the greatest chance of success. Alternatively, we might divide athletes into categories of genetically altered and genetically unadulterated.

Whatever we do, will people be more drawn to the nongenetically selected and unenhanced athletes with their slower times and decreased performance or to the enhanced superathletes who continually break records and expand our concepts of what humans can achieve?

Olympic champions, professional athletes, and other sports stars stand to reap enormous rewards and earn huge amounts of money from their athletic success, and they are often willing to take extraordinary measures to gain advantage. Some of these measures are dangerous, but compromising the health of athletes is the essence of some sports. Football, boxing, and free climbing are but a few examples. Athletes can always opt out; but competitive sports, like evolution, create arms-race environments, where each embrace of a particular advantage sets a new bar for other competitors to follow.

Given the massive financial and other rewards that come from being a top professional athlete, we can almost understand how people with these aspirations are willing to risk even their future well-being in pursuit of their dreams. But it's not just aspiring professional athletes willing to embrace new technologies that create these kinds of potential risks. Even average parents are doing the same to help their children realize their athletic hopes.

In the sports-obsessed United States, parents place their children in competitive sports programs when they are as young as four. Some twelve-year-olds have insanely competitive schedules almost equivalent to professional athletes. To serve this market of zealous parents, a large industry of direct-to-consumer genetic tests for athletic performance has emerged in recent years. Companies like Atlas Biomed, DNAFit, Genotek, Gonidio, and WeGene offer parents and others information on whether a tested person's genetic tests indicate single gene mutations believed to confer one athletic benefit or another. Although these predictions are today generally not that informative,[9] they will become far more operative as our understanding of the genome grows, creating new possibilities for parents.

As more genes associated with athletic performance are identified, parents will have the option to consider genetic predictors of athletic potential in one or another area when selecting which of their embryos to implant during IVF. Under current U.S. law, nothing would prevent clinics from offering, or parents from making, this type of determination.

Let's say parents selected an embryo to implant based on a predicted likelihood of the future child being a competitive sprinter or producing more than average hemoglobin, or simply tested a newborn for these genetic indications using a product from a direct-to-consumer providers. Would U.S. athletic associations and elite athletic training programs at least want to consider this information as one data point when deciding which young athletes to support, especially if competitor countries like China were already doing so? Jiaxue Gene, a private Chinese company based in Beijing reported on its website that it was already working with the Chinese government to screen children for sports-related genes. "The national sports teams and coaches have contacted Jiaxue Gene," the site noted, "to screen for students with the highest training potential through the company's genetic decoding technology."

In 2014, Uzbekistan became the first nation to announce it was integrating genetic testing into its national sports program. In conjunction with its national Olympic Committee and several sports federations, Uzbekistan's Academy of Sciences said it would test children for fifty genes believed to impact athletic potential to help identify possible future stars.[10] In August 2018, China's Ministry of Science and Technology announced that Chinese athletes aspiring to compete in the 2022 Winter Olympics would be required to have their genomes sequence and profiled for "speed, endurance, and explosive force" as one factor in an official selection process guided by "genetic markers." This type of testing today has a low probability of success because we still know relatively little about what genes do and because athletic success is such a complex mix of biological and environmental factors.

But given that the difference between a national hero world-champion sprinter and an also-ran is just a few fractions of a second, other countries and sports organizations will follow Uzbekistan's and China's model if there is even a hint of efficacy.

As advanced genetic testing becomes adopted by more countries to identify potential future Olympic stars, other countries will have a choice. If the leaders of sports associations believe genetic testing will not be able to drive success, that it won't matter, or that even if it works this type of testing violates the spirit of sportsmanship, they can choose to not genetically evaluate young athletes. If they believe genetic testing could have a significant impact on national athletic competition, they might ramp up efforts to do just that.

Once the hurdle of widespread genetic testing of athletes is cleared, the barriers to embryo selection, and ultimately limited genetic manipulation, to enhance athleticism will likely fall in some places for the exact same competitive reasons. People and leaders in the places where this Rubicon is not crossed will then similarly need to figure out how to respond. If they opt out, their national athletes may no longer be competitive in some sports.

Athletic competition is only one example of this human and technological arms race where advanced genetics will almost certainly be deployed, albeit in a relatively benign context. Who really cares, in the grand scheme of things, if the United States never wins another Olympic medal? But we all care a great deal that our children get good jobs and find career success, our national economies are strong, and our countries are able to defend themselves. The genetics of sports analogy isn't just about sports. It's also about life.

X

A truly miracle country, South Korea was in ruins following the devastation of the 1950–53 Korean War, with seemingly lesser prospects than the poorest African states. Through smart government policies, incredible hard work, and a relentless national devotion to education, however, the country grew from an annual per capita GDP of $64 in 1953 to $27,000 today, and from $41 million in total GDP to $1.4 trillion today—a phenomenal 31,000-fold increase.

Korean students today consistently rank among the highest in global comparative student assessments, and the country is among the most educated in the world. Because competition is fierce to ace the national university entrance exam, *suneung*, to get into the elite universities seen as stepping-stones to success, preparation starts young. In addition to the excellent schooling the government provides, 75 percent of Korean primary school children are enrolled by their parents in one of the hundred thousand or so cram schools around the country.

Based significantly on this pressure, youth anxiety and suicide rates in South Korea are among the highest in the world. In response, the Korean government in 2006 imposed a 10 p.m. curfew on cram schools because so many children were becoming chronically exhausted from working into the wee hours of morning at these schools. Despite government efforts to reduce the pressure on South Korea's children, the Korean education arms race continues.[11]

The race for competitive advantage in South Korea extends far beyond education. Physical beauty is treasured the world over, but Korea takes this to a whole new level. Even though the dangers of plastic surgery are well documented, South Korea has the highest per-capita rate of plastic surgery in the world. A BBC poll estimated that around half of all South Korean women in their twenties have had some type of plastic surgery. Korean parents often pay for these plastic surgeries as high school gradu-ation gifts to give their children an additional means of getting ahead.[12] "When you're nineteen, all the girls get plastic surgery, so if you don't do it, after a few years, your friends will all look better, but you will look like your unimproved you," a Korean college student told the *New Yorker* magazine.[13]

South Korean parents may be at one end of the spectrum in their willingness to even put their children's health at risk to gain competitive advantages, but they are not alone.

Getting kids into top middle and high schools in China often requires parents to use family connections, pay expensive school fees, and bribe teachers and school officials. The *Washington Post* reported in 2013 that "admission to a decent Beijing middle school often requires payments and bribes of upwards of $16,000, according to many parents. Six-figure sums are not unheard of."[14] Once they are in, the workload is enormous, and some children are in school seven days a week. According to a *Reuters* report, "Stiff competition for future jobs and ambitious parents mean long hours in the company of school books, not friends."[15]

Like the Korean *suneung*, preparation for the Chinese college entrance exam, *gaokao*, is brutal and stress inducing. High pressure can induce student stress and depression even among the younger students. A 2010 survey of nine- to twelve-year-olds in eastern China found that over 80 percent worried "a lot" about exams, 67 percent feared punishment by their teachers, and almost 75 percent feared physical punishment from their parents if they did not excel. A third exhibited regularly the telltale symptoms of extreme stress.[16]

American parents may not on average be pushing their children as hard as their Korean or Chinese counterparts, but the immense pressure on American high schoolers to do well on their SAT tests to boost their chances for college admission has sparked a multibillion-dollar test prep industry. A series of American books with titles like *The Pressured Child* and *The Over-Scheduled Child* argue the most aggressive parents are pushing the boundaries between helping and harming their kids.[17]

In the competitive race of life, these parents around the world are not wrong that advantages accrued after birth help children succeed. That's why they are also willing to go to such lengths to provide them, including by accessing new genetic tools.

Within the mix of the scores of new direct-to-consumer genetic tests being offered around the world, a small but growing number are being marketed to parents specifically for children. U.S. company BabyGenes offers parents information on about 170 genes with potential health implications. After sending in a mouth swab taken from their children, parents in various countries using these tests are told they can gain insights into their children's food tolerances, eating habits, sensitivity to second-hand smoke, susceptibility to addiction and hyperactivity, and whether the child is a morning or night person.[18]

MAP MY GENE'S 46 TALENTS AND TRAITS SPREAD ACROSS 8 DISTINCT CATEGORIES

PERSONALITY TRAITS:
Optimism, Risk taking, Persistence, Shyness, Composure, Split Personality, Hyper-Activeness, Depression, Impulsive, Mould-ability

EQ:
Affectionate, Faithfulness, Passion, Propensity for Teenage Romance, Sentimentality, Sociability, Self-Reflection, Self-Control

IQ:
Intelligence, Comprehension, Analytical Memory, Creativity, Reading Ability, Imagination

SPORTS:
Endurance, Sprint, Technique, Training Sensitivity, Tendency for Sports Injuries, Sport Psychology

ADDICTION:
Alcoholism, Smoking, General Addiction

PHYSICAL FITNESS:
Height, General Wellness, Obesity

ARTISTIC:
Performing, Music, Drawing, Dancing, Literature, Linguistic

OTHERS:
Sensitivity to Secondhand Smoke, Insensitivity to Secondhand Smoke

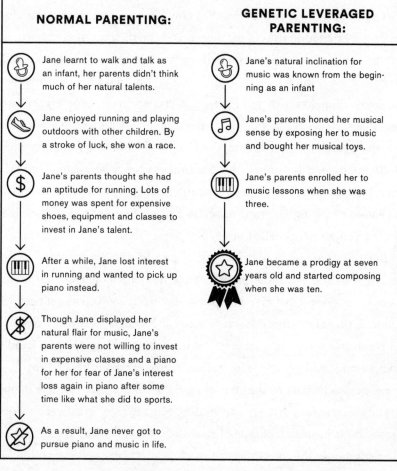

NORMAL PARENTING:

Jane learnt to walk and talk as an infant, her parents didn't think much of her natural talents.

Jane enjoyed running and playing outdoors with other children. By a stroke of luck, she won a race.

Jane's parents thought she had an aptitude for running. Lots of money was spent for expensive shoes, equipment and classes to invest in Jane's talent.

After a while, Jane lost interest in running and wanted to pick up piano instead.

Though Jane displayed her natural flair for music, Jane's parents were not willing to invest in expensive classes and a piano for her for fear of Jane's interest loss again in piano after some time like what she did to sports.

As a result, Jane never got to pursue piano and music in life.

GENETIC LEVERAGED PARENTING:

Jane's natural inclination for music was known from the beginning as an infant

Jane's parents honed her musical sense by exposing her to music and bought her musical toys.

Jane's parents enrolled her to music lessons when she was three.

Jane became a prodigy at seven years old and started composing when she was ten.

Source: Map My Gene, http://www.mapmygene.com/services/talent-gene-test/.

In China, dubious "health institutes" are now opening across the country claiming to predict the future strengths of children. One growing chain, Martime Gene, offers parents a genetic test allegedly identifying their children's talents in twenty different activities, including dancing, math, and sports. Some Chinese parents now pay about $1,500 for a genetic test called *myBabyGenome* that indicates 950 genes associated with disease risks, 200 with drug reactions, and 100 with physical and personality traits.[19]

The following images below from the website of the Malaysian company Map My Gene provide a good example of the type of messaging parents are increasingly receiving.

These types of genetic tests are, for now, far more useful for identifying single-gene mutation risks than for accurately predicting any general traits, but the competitive pressure on parents to access this still unproven and, for now at least, still unreliable technology is enormous. Recognizing this, some governments are stepping in to protect consumers. Believing the tests were not sufficiently accurate and that the general public could not handle too much access to genetic data, the U.S. Food and Drug Administration in 2013 temporarily banned the consumer genetics company 23andMe from providing predictive health information to customers about their likelihood of getting genetic diseases. In Europe, France and Germany ban direct-to-consumer genetic testing.[20]

But this level of caution will not always be entirely in order as the genetic tests become more predictive by providing polygenic scoring to more accurately predict even complex traits and as consumers get better at handling the information. Harvard medical geneticist Robert Green and his team have shown in a series of studies that regular people can handle even complicated genetic information if properly educated.[21] Providing complex genetic data to the general public would require massive public education, standard setting, and health-sector reform, but Green's work shows this is doable. This type of accessible testing, however, will only be the beginning of the consumer-driven adoption curve of genetic technologies, including by parents.

Would the subset of parents in Korea and elsewhere already willing to send their children to all-night cram schools be willing to select embryos optimized for the type of success to which the parents aspire? If they were already selecting embryos, might they be willing to genetically alter them to eliminate disease risk and, while doing that, make a few genetic tweaks that could enhance their future child's competitiveness in one area or another? If they are already deploying so much energy to give their children advantages accrued after birth, in other words, how big of a step will it really be for parents to want to give their children advantages accrued before birth?

The world's greatest engineers, computer programmers, and other technical specialists all benefit from a range of special advantages over their peers. Genetic aptitude may only be one element of this success. As in sports, grit resilience, emotional intelligence, personality, relationships, group dynamics, and luck also play a big role. But differences in any one of these, including genetic aptitude, could mean the difference between being—and being recognized and rewarded as—the world leader in many particular fields and not.

Because the prospect of an even partly genetically determined future is justifiably frightening to many people today, parents will rightly worry that making decisions about their future children's genetic makeup could diminish their children's sense of autonomy, agency, value, and free will. For these and other reasons, most parents will be able to partly or fully opt out of the genetic testing, assisted reproduction, genetic screening, embryo selection, and genetic engineering at every step along the way. But the choices to opt in, opt out, or do something in between will all come at a competitive cost parents, just like the national Olympic committees, will need to weigh.

Those parents who opt-in for full genetic optimization might, to give one example, ensure their children don't get certain diseases, live healthier longer, and have greater chances of excelling at a given task. If the opt-in benefits are great enough, we can imagine an accelerating curve of advantages. A generation of enhanced people could gain the advantages necessary to ensure that their children have greater access to the next generation of enhancements, and so on down the line. These people would also likely want to make sure their children only mated with partners with similar genetic alterations and that each generation was further genetically engineered so that the multigenerational impact of their enhancements would accrue.

Alternately, those parents who opt in might also be setting up their children for a lifetime of pain and suffering if the children aren't interested in performing the functions for which they have been engineered to excel. Because genetics won't be the only success factor, the child might also not be particularly good at what he or she has been optimized to do in the first place. There could also be a social backlash against enhanced people or some other unforeseen danger.

Those who opt out, on the other hand, might relegate their children to second-class status if their children can never match their genetically optimized or enhanced peers in some areas. The capacity difference between those opting in and those opting out, in this scenario, could grow consistently over time, potentially even creating two classes of people like H. G. Wells predicted in his 1895 novel *The Time Machine.*

Whatever happens, competition is likely to drive the process forward. And just like sports authorities and parents, states will also have significant competitive incentives to get on the genetic enhancement bandwagon.

X

When it became clear in 1944 that the allies would win the Second World War, American and Allied planners started imagining a better world that could emerge from the global ashes of destruction. Because they recognized fixed sovereignty and excessive nationalism as the cancers leading to the two world wars, these visionaries built institutions like the United Nations, the World Bank, and NATO to support new ideas of shared sovereignty and communal responsibility. They established concepts like transnational human rights and international law designed to temper the aggressive and all too often dangerous competition between states.

In many ways, these plans succeeded beyond their wildest dreams. American enlightened self-interest, backed by military might, provided the framework for growing peace and prosperity across much of the globe. In the decades following the war, the world experienced greater levels of economic growth, innovation, and improvements to the general well-being than in any previous period in human history, despite the decades-long Cold War between the United States and Soviet Union.

But countries, like people, exist in a highly competitive context resembling biological evolution, where the status quo never lasts forever. Just like biological organisms translate evolved capabilities into competitive advantages, countries translate the combination of the talents of their populations, governance structures, and natural resources into national power that can be deployed to educate more people, build better governments, get more resources, and gain advantages over others.

Although political theorist Francis Fukuyama declared that the "end of history" had arrived when it looked in 1989 as if liberal democracy had beaten out all other forms of government, this theory crashed into the rocks of the natural history of state competition.[22] After the 1991 collapse of the Soviet Union created a brief moment where U.S.-led globalism ascended, a new challenger was already emerging in the wings.

China has been a great civilization for over four thousand years that fell on hard times in recent centuries. Failing to modernize, it was defeated in a series of battles with European adversaries and Japan over the course of the nineteenth and early twentieth centuries, then suffered terrible destruction as Chinese nationalists battled the invading Japanese during World War II. After the Chinese Communist forces defeated the nationalists in a civil war and took control of the country in 1949, Mao Tse-tung declared that the Chinese people had "stood up" after so many years of foreign domination. But standing up under Mao would have brutal consequences for the Chinese people. In the twenty years following the revolution, Mao's policies devastated China and its people and wiped out its already small industrial and technological base. During the Cultural Revolution in the 1960s and '70s, many universities were closed, scientists were exiled to the countryside, and research came to a screeching halt.

Assuming power a couple of years after Mao's death in 1976, Deng Xiaoping recognized science and technology as "primary productive forces" needed to begin laying the foundation for China's growth. Universities were reopened. Scientists were rehabilitated.

But as China steadily became wealthier, it bristled at the perceived constraints imposed by American economic, political, and military dominance and agitated for a return to the exalted Middle Kingdom status China had once enjoyed. Achieving this aspiration required continued economic growth. China's leaders felt it also required a massive military buildup to protect China's access to raw materials and project power at home and abroad. More recently, President Xi Jinping articulated his aspiration for China to achieve its rightful place as a "global leader" in "comprehensive national strength and international influence" by 2050 by becoming a world leader in the technologies of the future. China set its sights on matching and then surpassing its greatest rival.

The United States has been the world's leader in science innovation for nearly a century and has reaped tremendous rewards from pioneering the jet, space, computer, information, biotechnology, genetics, and other scientific revolutions. America's technology leadership ensured that many of the world's leading companies in these sectors would be American ones. At the cusp of its breakout moment for a next generation of revolutionary technologies, including genetics, China's national obsession is not to miss the boat again.

Beijing's "Sputnik moment" arrived when DeepMind's AlphaGo algorithm trounced China's human champions in the game of Go. Already obsessed with harnessing their country's economic, military, and political might to challenge U.S. hegemony, China's leaders realized that becoming the world's leader in AI and other related technologies was the key to winning the struggle for the future across all areas of technology and national power.[23] Because advanced AI is so central to uncovering the secrets of the genome, this aspiration had major implications for the future of human genetic engineering.

Beijing's ambitious July 2017 Next Generation Artificial Intelligence Development Plan appeared to borrow liberally from the Obama administration's 2016 plan[24] but took the national security implications of AI leadership a significant step further. "AI has become a new focus of international competition," Beijing's document asserted, and China must "firmly seize the strategic initiative in the new stage of international competition in AI development, to create new competitive advantage, opening up the development of new space, and effectively protecting national security." The document set several goals for China: to be on par with the world's leading countries in AI by 2020; for AI to be its "primary driver" of industrial growth by 2025; and to become "the world's primary AI innovation center" and occupy "the commanding heights of AI technology" by 2030.[25]

The Chinese government's 2018–2020 Three-Year Action Plan for Promoting the Development of a New Generation of Artificial Intelligence Industry seeks to make China the leader in integrating AI in the health-care, robotics, manufacturing, and automobile and other sectors. The government recruited leading Chinese IT firms including Alibaba, Baidu, and Tencent to build national AI innovation platforms.[26] For the first time ever, more venture capital funding went to Chinese start-up AI companies in 2017 than to American ones.[27]

Doubling down on the national security implications of the AI revolution, the document called for the implementation of what it called a "military-civilian integration development strategy, to promote the formation of an all-element, multi-field, high efficiency AI military-civilian integration pattern."[28]

In the United States, by the time the Chinese plan was issued the Obama administration had been replaced by the Donald Trump administration. It took the new U.S. president a year and a half to appoint a science advisor to lead the White House Office of Science and Technology Policy (OSTP), and even by then a majority of the staff positions in the OSTP had not been filled.[29] The president's proposed 2018 budget called for major cuts at the National Institutes of Health, the National Oceanographic and Atmospheric Administration, the National Science Foundation, and other federal agencies involved with AI research and science more generally. Immigration visas, including for exceptionally talented specialists, were restricted. Among the few recognitions by the administration of the importance of AI was the creation of a White House Select Committee on Artificial Intelligence in May 2018 , but little seemed to come from the effort.[30]

With the U.S. government less focused on supporting the technologies of the future, China is vying aggressively to assume leadership in unlocking the secrets of the genome and ushering in the new era of precision medicine to help realize the country's broader strategic aspirations.

Evidence of China's commitment to winning this race in advanced genetics and personalized medicine is everywhere. China's recently announced plan to establish national leadership in precision medicine,[31] for example, dwarfed the much smaller initiative pledged by the Obama administration, later undermined by President Trump.* Although the U.S. company Illumina remains the leader in building advanced sequencing machines, China is quickly becoming the dominant power in bringing together the massive data sets that will drive the next phases of understanding how genes function.

U.S. research capabilities remain the best in the world, but Chinese authors' publications in the world's leading medical journals have doubled in recent years, and the number of U.S. patents granted to Chinese inventors and companies is increasing at nearly 30 percent annually, far faster than their American counterparts.[32] Chinese research and development spending has increased by an astounding average of 15 percent a year for most of the past two decades and is now second largest in the world. It is still less than the United States but more than all the European Union countries combined. China also awards more science and engineering PhDs than any other country.[33]

With their massive scientific, investment, and industrial bases, the United States and China are increasingly competing in a two-horse race to be the leading economy and scientific powerhouse of the future. This competition will impact every advanced technology influenced and revolutionized by AI, genomics in many ways foremost among them. "In the age of AI, a U.S.-China duopoly is not just inevitable," Kai-Fu Lee, founder of Beijing-based technology investment firm Sinovation Ventures, and former Microsoft and Google executive, recently asserted. "It has already arrived."[34] As researchers Eleonore Pauwels and Pratima Vidyarthi write, "Increasingly, the U.S.-China relationship will not be defined by the ownership of 20th century manufacturing industries but by a race in genetic and computing innovation that will drive the economy of the future."[35]

---

*   In a snub to the Trump administration, Congress rejected requested cuts to the precision medicine initiative and increased the program's budget in 2017.

Some of this competition will be win-win, where humanity at large benefits from a faster rate of progress realized through competition. Some of it will be zero-sum, where some societies, companies, and individuals will lose power, market share, and perhaps even autonomy as others gain them. Because the wealth and power that will accrue to those countries, companies, and people who lead these revolutions is immense and unpredictable, the race is on. The companies and countries that crack the code of diseases like cancer and thousands of others and gain the ability to understand and potentially alter other human traits will be not just the biology versions of Google and Alibaba but have the potential to be more equivalent in their power and influence to the thirteenth-century Mongols, the nineteenth-century British, or the twentieth-century Americans.

As these two great ecosystems of innovation compete, each will have a unique set of strengths and weaknesses that will together determine how this rivalry plays out and impacts the application of genetic technologies.

China, for example, will have larger genetic data sets, because its population is larger, it is collecting genetic samples much more aggressively, and because data privacy regulations are far weaker than in the United States or Europe. America and Europe's higher levels of privacy protection for personal data could help maintain public support for genetics research and applications or could become an Achilles' heel if it keeps American and European researchers from accessing genetic data sets as robust as that of their Chinese counterparts.[36]

As this rivalry plays out, new national champion companies will emerge, the life sciences equivalents of the American FAAMG technology behemoths (Facebook, Apple, Amazon, Microsoft, and Google) and Chinese BATs (Baidu, Alibaba, Tencent). Already, American companies like IBM and new Chinese companies like iCarbonX are positioning themselves to assume this mantle.

Founded by former BGI CEO Wang Jun in October 2015, iCarbonX seeks "to build an ecosystem of digital life based on a combination of an individual's biological, behavioral and psychological data, the Internet and artificial intelligence."[37] By combining comprehensive biological, patient-generated data with AI technology, it plans to help consumers better understand the medical, behavioral, and environmental factors in their lives to optimize their health and to help companies use genetic data to optimize their products and services. Wang Jun's plan is to ultimately build a "predictive digital avatar" of hundreds of millions of customers,[38] allowing them to pass from sequencing to fully digitizing themselves.[39] "We can digitalize everyone's life information," Wang Jun says, "interpret the data, find more valuable law of life, and thus enhance the quality of people's lives."[40] With Tencent and the private equity behemoth Sequoia capital as early investors, by 2017 iCarbonX quickly shot up to a $1 billion valuation, becoming China's first biotechnology unicorn.[41]

This convergence of competitive pressures on the individual, business, and societal levels will push the boundaries of how genetic tools are utilized and confer an increasingly broad set of advantages on the people, companies, and countries who optimize their development and application. But while all of these competitive pressures drive the adoption of genetic technologies forward, the same types of competitive pressures also have the significant potential to drive conflict within and between communities and states.

X

Imagine you are the leader of a society that has chosen to opt out of the genetic arms race by banning embryo selection and genetic alteration. Because your country is progressive enough to make a collective decision like this, parents desiring these services are free to go elsewhere to get what they want. But preventing the genetic alteration of your population by definition requires both restricting genetic enhancement at home and enhanced people or expectant mothers carrying genetically altered embryos from entering your country.

To protect the genetic integrity of your populations and keep genetically enhanced people out, you would need to perform genetic tests on all people entering the country. But there would likely be no way of knowing whether a person had been genetically enhanced without knowledge of their genetic baseline—their genome prior to any changes. For those few people for whom genetic information from the moment a few days after their conception is available, their former and current genetics could be compared. Everyone not able to provide baseline genetic information might be banned from entering the country or threatened with long jail terms for procreating with a citizen of it.

To prevent women from going abroad to have genetically engineered embryos implanted, pregnancy tests would need to be performed on all women of fertile age coming into your country. Prenatal blood tests would then need to be performed on the pregnant women to try to guess if the embryos had been manipulated in some way. Even with a list of the most fashionable genetic alterations,* this would be all but impossible. To be effective, these types of blood and prenatal tests would probably need to be accompanied by a polygraph asking pregnant women if they are carrying a genetically enhanced embryo.

If someone already in the country was identified as enhanced, what penalties could possibly be meted out? Even if enhanced people were stripped of their citizenship and exiled for giving birth to a genetically enhanced person, their children would also need to be imprisoned, banned from procreating, or exiled. Enforcing any of this would require building the oversight machinery of the most totalitarian, intrusive, abusive, and downright odious police state with the ability to track peoples' movements and continually monitor their biology and that of their children.

But let's say your country has done all this and become a preserve of nongenetically enhanced people. We've already seen why different states will adopt advanced genetic engineering technologies at different rates based on the significant historical, cultural, and structural differences between them. Imagine you are assessing your country's options in a world

---

* This would be something like the virus-scanner programs on computers, which continually generate lists of potential viruses to defend against.

where your country has opted out but other countries are moving forward with human genetic enhancement. Here are your general choices:

Option 1: You recognize that your country has made a moral decision based on your collective values and accept facing the consequences, even if this means your country will gradually lose its competitive advantage and future generations will be less healthy, live shorter lives, and have fewer superstars of various sorts. You sit tight in your belief you've made the right choice. With *schadenfreude* in your heart, you hope your national decision will give you a competitive advantage if and when human genetic enhancement proves to be less beneficial and more dangerous than initially believed. Because your country has taken such a strong and principled stand on human genetic engineering, you feel duty-bound to protect this ban against encroachment. You are a progressive in your heart but recognize you'll need some trappings of a police state to maintain your country's genetic purity. How is it, you ask yourself late at night, that an idealist like you is starting to adopt the language of Nazism?

Option 2: You try to hold the line and support your national decision but feel the pressure growing. Many of your most talented people are leaving the country to get the genetic enhancement services they want. Your unenhanced aspiring Olympic athletes and advanced coders are becoming community organizers, yogis, and nurses instead, pursuing careers that don't require competition with their enhanced counterparts. Parents are having second thoughts about your ban as they hear about kids in other countries who are immune to genetic diseases, doing better on IQ tests, and achieving all sorts of seemingly superhuman feats. Your military is worried your future soldiers will be at a disadvantage compared to their genetically enhanced adversaries. The leaders of your national space program tell you that your unenhanced astronauts will, unlike their enhanced counterparts from other countries, not be able to withstand the radiation exposure and bone density loss of extended space travel. Opting out is seeming less appealing an option. You need a face-saving alternative. You call for a national referendum. After a heated debate, you cast your vote to opt in.

Option 3: You see the benefits of genetic enhancement, but your citizens still believe meddling with the human genome and rewriting biology is a form of hubris likely to end badly. As a matter of principle, you recognize that societies, like people, are diverse and don't begrudge the many other choices different societies make in all sorts of areas. But this is different. If other societies genetically enhance their populations and yours doesn't, you may not just be at a competitive disadvantage in the future. You may not be able to protect your population from the very thing they have so adamantly opposed. Just like genetically modified crops spread into adjacent fields and gene drived mosquitoes spread across national boundaries, there will be really no way to protect your population from inheriting what you see as unnatural genetic modifications unless other countries can be prevented from allowing the most egregious modifications. Your only option is not just for your country to opt out but to define, promote, and seek to enforce limits on genetic enhancement for all countries to follow. You ask your top advisers how you can make this happen.

First on their list is trying to use your national powers of persuasion to convince people and countries around the world that the downsides of human genetic enhancement outweigh the benefits. But what are the chances of your being able to convince the whole world to buy into your pessimism, particularly when other societies are enthusiastically racing forward into the genetics age?

Second, you can try to build an alliance of like-minded states to collectively pressure other countries to limit genetic enhancement. Getting an enforceable global treaty to limit genetic enhancement is an appealing option, but it's difficult to do. Most global leaders agree that human-induced climate change is threatening the livability of our planet, but we've not been able to get an enforceable global treaty to turn things around. Could a global effort limiting a technology many people and other states support be more effective than the high-profile efforts to limit climate change?

Third, you identify the enhancing countries you are most concerned about and, if you have the power and influence to do it, try to stop them to set an example. One Central Asian country in particular has become a hub for highly aggressive genetic alterations of preimplanted embryos designed to create superhuman capabilities. Parents are sending their frozen eggs and sperm, or skin grafts and blood samples from which these sex cells are being generated, to this country for embryo selection, embryo mating, and genetic enhancement.[42*] For the Central Asian country, building this industry is seen as a moral imperative, a great business opportunity, and a strategic boon. You ask them nicely to stop. They refuse.

Perhaps you try getting a group of countries to impose travel, economic, or other sanctions on the offending country. If none of these approaches work, are you willing to use military force to stop the genetic alteration of the human species? It's certainly one option on the list.

Over the course of the twentieth century, an estimated 170 countries were invaded by others for a whole host of reasons, ranging from outright theft to ideological differences to pre-emption of a wide range of perceived threats.[43] Is it so outlandish to believe that countries in the future might resort to military force to prevent other countries from altering the shared genetic code of humanity? Many countries have been invaded for far less.

Military force would be an option if advanced genetic enhancement were only being carried out in a relatively weak country or even in international waters or space. But what happens if a powerful country like China takes the lead in deploying advanced genetic and other technology to enhance the capabilities of its populations while another country, say the United States, has entirely opted out for political and other reasons? Would the United States and China be willing to use as much force over the potential transformation of our species as they are now threatening over a few contested reefs in the middle of the South China Sea?[44]

If all of these types of competitive pressures on the personal, communal, and national levels were rare in our human experience, an argument could be made that they could be avoided in the context of the genetics

---

* This is not unthinkable. A 2014 *New York Times* article described a Chinese parent who sought to have six children born from U.S.-based surrogates to then choose the "pick of the litter" and put the others up for adoption.

revolution. But because competition has been at the very core of our evolutionary process for almost four billion years, the overwhelming odds are
that these same drivers will push us, unevenly but collectively, into our
brave new world of increasingly sophisticated human genetic engineering.

Both the competitive pressures pushing human genetic engineering
forward and the potential conflict scenarios this competition is likely to
spark are very real. If we do nothing to apply our best values to influence
how the genetic revolution plays out, we will place ourselves on a path to
conflict. Avoiding worst-case scenarios will require our species to come
together as never before to figure how the benefits of revolutionary genetic
technologies might be optimized and the dangers minimized.

The good news is that we've tried to do this kind of thing before. The
bad news is that we've never fully succeeded.

Chapter 11

The Future of Humanity

The early nuclear scientists understood both the creative and destructive potential of their work. "The splitting of the atomic bomb has changed everything save our mode of thinking," Einstein wrote after the United States dropped atomic bombs on Hiroshima and Nagasaki and the Cold War began, "and thus we drift toward an unparalleled catastrophe." Just as nuclear power could help us build a better future, nuclear weapons could destroy us.

American postwar leaders also recognized this dual promise and peril of nuclear power. Even though the United States had a monopoly on atomic weapons at the end of the war, some American officials argued the U.S. should share its nuclear secrets with the Soviets to prevent a dangerous arms race. Others, like State Department strategy guru George Kennan, believed the United States should leverage its atomic monopoly to resist Soviet aggression.

After a U.S. proposal for international control over nuclear materials, global inspections of all nuclear sites around the world, and the active sharing of atomic energy technologies for peaceful means was rejected by the Soviets, the USSR tested its first atomic device in August 1949. The nuclear arms race was on. The British detonated their first nuclear weapon three years later, in 1952, the French in 1960, and the Chinese in 1964. Our world was fast becoming a much more dangerous place.

Global efforts to balance the legitimate desire for nuclear energy with the existential danger of an unbounded nuclear arms race took a small step forward with negotiations in the 1960s to establish a nuclear arms treaty. Ratified in 1970, the Treaty on the Non-Proliferation of Nuclear Weapons, or NPT, did two critical things. First, it established standards for nonproliferation in the five countries by then permitted to possess nuclear weapons: Britain, China, France, the United States, and the USSR. Second, it created a set of incentives to encourage other states to refrain from developing or acquiring nuclear weapons in exchange for a promise to help them develop nuclear energy for peaceful purposes.

Since ratification, the NPT's impact has been imperfect at best. On the positive side, the world hasn't seen the nuclear arms free-for-all many feared. The acquisition of nuclear weapons by nonnuclear states also remains an important, if weakening, taboo protecting humanity. On the other hand, the United States and Russia today have the nuclear weapons to blow up the planet many times over; both Ukraine and Libya were invaded after giving up their nuclear weapons programs; Israel, India, Pakistan, and North Korea have acquired nukes outside the NPT; and the danger of a global nuclear arms breakout is real and growing. It's not difficult to imagine a regime for nuclear arms reduction that might have worked better. But we're all still much better off even with the flawed system we have.

When I started thinking many years ago about how the worst dangers of the genetics revolution might be prevented, I kept coming back to the nuclear arms example.

Like the nuclear arms race, an international competition in the field of genetics—a genetic arms race—has enormous potential to either improve people's lives or do them harm. Both represent technological capabilities developed in more advanced countries that become desirable and ultimately accessible the world over. Having nuclear weapons in one country may empower that country, but having nuclear weapons at all or in multiple countries threatens us all. Likewise, human genetic engineering has the potential to significantly help individual humans and countries, but an uncontrolled genetic arms race could harm humanity.

At first blush, the idea of regulating the miraculous genetic technologies that researchers around the world developed with the most noble intentions feels somehow wrong. The nuclear era began at Hiroshima and Nagasaki. The worst case was realized before the benefits became apparent. The genetic era is beginning in research labs, with scientists finding cures for our most debilitating diseases and IVF clinics helping loving parents make healthy children. The dangers of human genetic engineering remain more hypothetical and in the future.

But as tempting as it may seem to some libertarians and transhumanists to keep governments out of science's way in these benign early days of the genetics revolution, that is not the right approach. It will simply be too dangerous for everyone if some of us start remaking the biological code of life on earth without any common rules of the road. This will become even more the case as access to the genetic revolution's most powerful tools are democratized.

Most of the NPT signatory states gave up the right to possess nuclear weapons in exchange for help developing their civilian nuclear industries because they recognized that a world where every country had nuclear weapons would be an inherently more dangerous world. If we followed the same model for human genetic engineering, countries would need to feel they were giving up the possibility of using unlimited and unrestrained gene editing on humans in exchange for assurances their countries, and the world at large, would be better off by developing these technologies for the common good. This may sound like a simple proposition but it is not, particularly because, as we've seen, the aims of different people, groups, and countries are so diverse.

Like with the nuclear revolution, the early pioneers of the genetic revolution are professors and researchers at important institutions. A few have already or will soon win Nobel Prizes. The people applying their work will be well-trained local doctors and technicians in IVF clinics and university, hospital, and corporate laboratories and clinics around the world. Someday in the not-distant future, however, the next generation of today's do-it-yourself, or *DIY*, biohackers—people doing biological work outside of professional laboratories—will be able to make meaningful alterations to living organisms, including future humans, on their own.

X

The biohacking movement is exploding around the world. In a 2005 *Wired* article, scientist Rob Carlson outlined how he built a powerful genetic-engineering lab in his garage spending under a thousand dollars on eBay.[1] Carlson wasn't birthing a genetically altered Frankenstein or giving his dog sonar but foreshadowing our decentralized world to come. Today there are more than fifty DIYbio community spaces in the United States, sixty in Europe, twenty-two in Asia, twelve in Canada, sixteen in Latin America, four in Australia and New Zealand, and a few in Africa.[2] Biohackers are largely unregulated and deploying increasingly powerful technologies, and they will, over time, significantly decentralize the ways and places where the genetic engineering of life happens.

In these still-early days, some of the biohacker applications—like the use of genome sequencing to determine whose dog is responsible for the poop on your lawn—are amusing. Others, like making cheap batches of "homebrew" synthetic insulin, are potentially more useful.[3] Soon, however, DIY biologists will have access to almost unimaginably power-ful and inexpensive tools, like desktop genome printers that can combine easily acquired genetic fragments to recreate life. As this happens, these DIY biohackers will become to the scientific establishment what home computer hobbyists like Steve Wozniak ended up being to the established mainframe-computer companies like IBM—seemingly irrelevant outsid-ers who proved a heck of a lot more significant than they first appeared.[4]

As more and different types of people have access to advanced genetic-enhancement capabilities, the potential upside benefit of greater innova-tion and the downside risk of abuse will both grow.

More people using increasingly powerful technologies to solve more complicated problems will lead to greater innovation, which will expand the capacity of genetic engineering to enhance our lives faster than most of us can imagine. "The most remarkable fact about human genetic engineering today," Siddhartha Mukherjee writes, "is not how far out of reach it is, but how perilously, tantalizingly near."[5]

At the same time, this proliferation of knowledge and capability will bring real dangers. In addition to the possibilities for international conflict stemming from the uneven application of genetic technologies, gene editing could be used by states, terrorist groups, or individuals to create deadly pathogens with the potential to kill millions.[6] Recently, for example, a team of researchers at Canada's University of Alberta for around $100,000 recreated from DNA fragments the horsepox virus, a dangerous relative of smallpox, in an effort to develop a vaccine to a related modern disease.[7] Doing something like this would have been unimaginable or prohibitively expensive just a decade ago, but now the technology is not a big deal and the costs are dropping precipitously.

These changes have the potential to bring a whole new set of actors into the genetic-engineering world and, with them, both new opportunities and dangers. Just like responsible scientists are weighing options for using targeted gene drives to push genetic changes through animal populations to do things like eliminate Lyme disease and malaria, rogue states, terrorists, or even well-intentioned scientists now have the ability to introduce just a small number of genetically altered organisms that could crash entire ecosystems.[8]

Recognizing this type of danger, the U.S. intelligence community for the first time included gene editing as a potential weapon of mass destruction threat in its 2016 *Worldwide Threat Assessment*. James Clapper, then U.S. director of national intelligence, wrote:

> Research in genome editing conducted by countries with different regulatory or ethical standards than those of Western countries probably increases the risk of the creation of potentially harmful biological agents or products. Given the broad distribution, low cost, and accelerated pace of development of this dual-use technology, its deliberate or unintentional misuse might lead to far-reaching economic and national

security implications.[9]

As Clapper recognized, maliciously or inadvertently deployed technologies have the potential to cause tremendous damage. "The very nature of life and how people love and hate," the U.S. National Intelligence Council wrote in its 2017 *Global Trends* report, "is likely to be challenged by major technological advances in understanding and efforts to manipulate human anatomy, which will spark strong divisions between people, country and regions."[10] In June 2018, the U.S. National Academies of Sciences, Engineering, and Medicine released a high-level report, *Biodefense in the Age of Synthetic Biology*, warning that inexpensive and accessible new synthetic tools could be used by terrorists to create deadly and potentially fast-spreading pathogens.[11]

We already see how different countries have different national approaches toward regulating, or in some case not regulating, genetic engineering technologies. No matter how well individual countries regulate themselves, addressing the potential dangers of the misuse of genetic technologies will also require meaningful regulation on an international level. But if creating national restrictions on how these technologies can be used is hard, doing this on an international level is even harder.

X

"This spear thing is great," one of our more thoughtful early ancestors must have once said, "but I'm concerned that left unchecked we could use it to kill the other people in our own tribe."

The dirt-crusted humans crouching around the fire gnawing at bones nodded their assent. "Mmmm, good," they grunted. "Don't kill others in our own tribe. Good."

But our thoughtful early ancestor, a visionary, had an even more enlightened idea.

"This isn't just about not killing people in our own tribe; we need to prevent deadly conflict with the other tribes. We have these spears today, but let's get real. How hard is it to make these things? We just sharpened the end of a stick. Once they see what we've figured out, the other tribes are going to make spears just like ours. Then what happens? Maybe they kill us. Why don't we bring together the neighboring tribes now and all agree we'll only use these spears to hunt food in the safest and most environmentally sustainable manner possible and not harm each other?"

Cue sound of crickets.

"I've got another idea," a scrawny man in a faded, second-hand loincloth chimed in. "Let's use our spears to hunt animals *and* steal all the other tribe's stuff. Then we'll have more food *and* more stuff."

The heads started nodding vigorously. "Mmmm. Good idea."

Creating regulations across group borders has always been a tall order. It is a decent bet that someone like our sagacious ancestor warned of a looming threat and proposed an international regulatory effort at the start of every technological revolution and to largely no avail. The industrial age brought tremendous power and wealth that humans used to kill each other at an industrial scale. The internet connected the globe in novel ways but opened the door to new levels of abuse, manipulation, aggression, and societal control. As Richard Clarke and R. P. Eddy write in their important 2017 book *Warnings*, the Cassandras raising the alarm about not-yet-realized dangers are often ignored.[12]

It's ironic: If someone suggests creating international regulations now for a potential future danger, many will say it's unnecessary and not urgent. But if a system isn't put in place, it can be too late to walk things back when a threat actually appears in the future. Exactly because managing any powerful technology is so difficult, we need to look now at the better examples of how we've dealt with other revolutionary technologies, like atomic energy, on an international level for ideas about how we might find the right balance between promoting the great benefits of human genetic engineering and minimizing the worse potential dangers.

Imagine what would have happened if we'd thrown up our hands in 1945 and said nuclear weapons proliferation was inevitable, so we may as well let it happen. Imagine how much worse things would be if we just accepted the destruction of our environment without fighting to save our planet and find alternative energy sources.* Internationally negotiated treaties on the use of chemical and biological weapons have not been enforced perfectly—or in the case of Syria, at all—but they are still far better than the alternative. We build cars that can go almost eight hundred miles per hour but collectively choose to impose speed limits that balance people's desire to get where they are going quickly with a societal need for safety.

To find the right speed limit for human genetic enhancement, we will need to find a way to harmonize approaches to human genetic engineering around the world before the consequences of communal and national differences tear us apart. We must find a path forward that avoids the unrestricted fantasy of transhumanism and the dehumanizing legacy of eugenics.

As a start, we'll need to figure out what we, collectively as a species, do and don't want to restrict.

There is nothing inherently wrong with human genetic engineering or enhancement. If our species was made perfectly, we would not have all the glitches in our biological software that cause people to suffer disease, die young, or not realize their potential in any number of areas. Even those who believe in god might still imagine their maker created the cosmos as something more than wallpaper for the human sky and wants us to figure out how to amend our biology to make exploring and inhabiting distant solar systems possible, particularly after our planet becomes uninhabitable.

It may end up being that limiting human genetic engineering in a

---

*   Global efforts to slow climate change show how difficult it will be to build norms and rules around human genetic engineering. It took decades for governments around the world to recognize that human-driven climate change threatened to wreak havoc on the environment and make large parts of the globe virtually uninhabitable for humans. But in spite of the overwhelming consensus among scientists and most world leaders that climate change is real, man-made, and an existential threat to our planet and selves, little progress has been made. Leaders at the 2017 meeting on the Paris Accord were only able to agree that each country would do its best without agreeing to any binding limits on greenhouse gas emissions. Even that wasn't sustainable when U.S. President Donald Trump later withdrew the United States from the agreement.

uniform way around the world could become a collective evolutionary disadvantage. For all we know, some natural or man-made disaster like a deadly virus or nuclear war is just around the corner that we won't be able to survive with the biology we've evolved to date. It could alternatively be that altering our evolutionary course could make us less diverse and competitive in a world we can't now foresee. But while the argument for pushing genetic technologies forward is a compelling one, not drawing any lines is a recipe for disaster.

If the danger of a genetic free-for-all is likely to be conflict within and between societies, we must find the best possible alternative. What our few examples of past success tell us is that smart international regulation must be a key component of our approach.

〼

For the past few years, top scientists have been meeting repeatedly to propose a best way forward on human genetic engineering. Some observers have warned that by getting involved at too early a stage, governments might do more harm than good. Others have suggested that self-regulation by the scientific community should be enough for now.[13] Although the scientific community has done an admirable job laying out a prudent path for using genetic technologies wisely,[14] this self-regulatory approach is not enough. The stakes for our species are simply too high to let human genetic engineering go completely unregulated. As tempting as it is for some of us to support letting a hundred flowers of different regulatory and nonregulatory approaches bloom, we must strive for some sort of global harmonization around how far we as a species will go evolving ourselves.

At its best, global governance balances the complex and often conflicting interests of diverse countries and groups. Just like the Treaty on Non-Proliferation of Nuclear Weapons balanced the needs of nuclear and non-nuclear states, any effective global regulatory structure on human genetic engineering would need to balance the interests of all countries.

Those who see the unfettered genetic alteration of their future children as a basic right, and perhaps even an obligation, will need to find at least some common ground with those who see it as an affront to human dignity. Finding this balance is easier said than done. The challenges begin at first principles.

Although philosophers have struggled for millennia to define human rights and responsibilities, the *Universal Declaration of Human Rights*, created in the aftermath of the two world wars and Nazi atrocities, provides a common standard for "the inherent dignity and…equal and inalienable rights of all members of the human family." In its first clause, the declaration states: "All human beings are born free and equal in dignity and rights." That's where things get complicated.

What does the world *born* mean in this context? Does it apply to an early-stage embryo genetically altered prior to implantation? Changing the word *born* to *conceived* not only limits women's reproductive rights but also means that each of the thousand fertilized eggs in a dish derived from a woman's induced stem cells have the same rights as her ten-year-old child. The issues, in other words, are extremely complex.

If we were able to reach consensus on the rights being protected and how to protect them, however, a globally harmonized regulatory structure would have a lot of benefits. In addition to reducing the possibility of both conflict and dehumanizing human experimentation, it could also facilitate international cooperation, reduce costs of regulatory compliance, and foster an environment of global collaboration for the common good.

With these types of goals in mind, a few—mostly misguided—efforts have been made over recent years to reach some type of international consensus on the way forward.

After the UNESCO International Bioethics Committee asserted that "the human genome must be preserved as common heritage of humanity,"[15] the 1997 UNESCO *Declaration on the Human Genome and Human Rights* prohibited "practices which are contrary to human dignity, such as reproductive cloning of human beings." The same year, the Council of Europe opened for signature its *Convention on Human Rights and Dignity* *with Regard to Biomedicine,* which asserted that interventions aimed at modifying the human genome can only be undertaken "for preventive, diagnostic or therapeutic purposes and only if its aim is not to introduce any modification in the genome of any descendants."*

In February 2002, the United Nations Ad Hoc Committee on an International Convention against the Reproductive Cloning of Human Beings began negotiations intended to lead to a binding treaty. The nonbinding General Assembly resolution adopted in March 2005 by a vote of 84 in favor, 34 against, and 37 abstentions, called on members to "prohibit all forms of human cloning inasmuch as they are incompatible with human dignity and the protection of human life."[16] A 2015 UNESCO meeting report called for a moratorium on germline modification because, it argued, making heritable changes to the human genome would "jeopardize the inherent and therefore equal dignity of all human beings and renew eugenics."[17] At the end of 2015, the Council of Europe committee responsible for reviewing the 1997 convention issued a statement backtracking from the earlier document, referring to the convention's core principles as merely reference points for a future debate.[18]

Not only are none of these international agreements and documents legally binding, they don't even represent a global consensus on the way forward. The countries with the most to gain from, the greatest hopes for, and the greatest cultural acceptance of this science and its more aggressive applications mostly didn't sign on because they were reluctant to have their activities limited by others with less skin in the game.

Even more importantly, seeking to restrict "any modification in the genome of any descendants" and thinking of the human genome as a sacred "common heritage of humanity" was an overly simplistic approach two decades ago when few people then could have imagined the clinical miracles that would later arrive. By licensing in 2018 the

---

* As of March 2018, 35 of the 47 Council of Europe member countries had signed the convention, but only 29 had ratified, and not all of these passed domestic implementation laws.

first clinics to perform mitochondrial transfer between embryos, the United Kingdom, among the most thoughtfully regulated countries in the world for reproductive technologies, was not a rogue actor defiling the "common heritage of humanity" but a medical and humanitarian pioneer. The so-called "three-parent babies" born from this process are not outlaws but, like Louise Brown in the 1970s, healthy forerunners of the world to come.

With all the new applications in the pipeline for genetically altering ourselves and our future children to eliminate and reduce the risk of disease, avoiding "any modification in the genome of any descendants" starts looking less like a humanitarian gesture and more like an investment in future danger. Seeking to protect the human genome as a "common heritage of humanity," if taken literally, becomes an argument for banning sexual reproduction and requiring people to only reproduce by cloning. Even if Darwin's principles of evolution in humans are hacked, they too will continue to evolve—just by different means. Evolution itself is evolving.

But if efforts to create international consensus around human genetic engineering have so far failed, we must still find a way to do better going forward in order to optimize the future of our species.

Like the NPT, an international regime on human genetic engineering would have the tough, dual role as both an enabler of responsible science applied for the common good and an enforcer of a limited restrictions on how far these activities can go.

Finding this balance would be incredibly difficult to negotiate. Draw the line too permissively and the opponents will likely revolt. Draw it too restrictively and proponents will find an alternate way to get the services they want, using underground clinics, other countries, or extra-national environments like the high seas or, someday, space. Draw the line down the middle and any regulation is likely to be too vague to have any real meaning. Fail to establish one global standard for human genetic engineering and this revolutionary set of technologies will become subsumed in an international genetic arms race.

An agreement would need to neither offend the sensibilities of powerful constituencies deeply uncomfortable with the concept of human germline engineering nor impede the development of new generations of knowledge and its application on which trillions of dollars of commerce; the competitiveness of individuals, companies, and countries; and the well-being of future generations depends. Within these narrow bounds, any standard would need to be extremely permissive, limiting only the most obvious violations of human dignity.

Because governments don't always perfectly represent everybody, we would also need to figure out whose interests would be represented in negotiations, how these interests would be organized, what type of enforcement mechanism could possibly be deployed, and how the arrangement could be used to advance the best science rather than impede it.

At this stage, probably the only restrictions that could be agreed upon would be a common definition of what constitutes a redline beyond which human genetic engineers should not pass. This would include potential behaviors most everyone would agree are reprehensible, such as dangerous and dehumanizing human experimentation, creating excessive mixtures of human and animal genes, and engineering humans with new and radical types of synthetic traits that challenge human identity. The average fertility clinic or university lab is not remotely considering any of these today—just like few university physics departments and nuclear power stations are also creating nuclear weapons.

Because both the underlying science and our cultural acceptance of what new genetic applications seem normal, healthy, and advantageous will change over time, any international standard would need to be extremely flexible and to be renegotiated regularly to accommodate new technological and medical developments.

If, against all odds, we as a species succeed in creating this kind of preliminary global regulatory structure, we will still need to figure out how to balance the carrot of enticement and the stick of enforcement. As with the NPT, I can imagine the more advanced countries offering to help bring others into the genetic and biotechnology age in exchange for commitments by the less developed countries to join an international regulatory body and abide by agreed ethical principles.

Even if this happens, there will always be a market for some kinds of extra alterations and enhancements banned in most parts of the world. Perhaps aggressive parents decide to give their children dangerous predatory instincts. Perhaps a given country decides to genetically engineer a subclass of its people to be superhuman masters or docile followers or fierce killers at the behest of the state. Perhaps companies will set up extreme genetic enhancement centers in international waters. What will happen then?

At a 2005 global summit, heads of state from around the world endorsed the principle that states have a "responsibility to protect" their citizens which, when violated through human rights violations and other abuses, shifts to the international community.[19] Although the "responsibility to protect" is not legally binding, it was used to justify military interventions in places like Iraq, Libya, and Yemen with mixed results at best.

For only the most egregious violations of agreed international norms on genetic engineering, the international community could conceivably fashion an escalating level of pressures, ranging from economic sanctions to military intervention, to change the ways of a given violator. Given the poor track record of many forcible "humanitarian" interventions, it may seem downright crazy to suggest that countries might someday justifiably intervene militarily to stop a country from genetically engineering its population in some monstrous way that massively violated an internationally agreed, ethical redline. From the perspective of today, it probably *is* crazy. But if we as a species determine that some future alterations of our fellow human are far beyond the pale of acceptability, some ultimate enforcement mechanism might someday be necessary.

Given all of these challenges, establishing and enforcing any sort of global restrictions on human genetic enhancement is a very tall order for which we collectively are not close to being ready. Today's mismatch between what the science can and will soon be able to achieve and how poorly people understand and are prepared for it is creating an extremely dangerous public tinderbox that must be addressed, first and foremost, through public education and engagement.[20]*

A first, immediate step in this direction is to help every country develop its own national public-education program, bioethics commission, and regulatory framework for human genetic engineering that applies its own traditions, values, and interests.

There are many excellent models from around the world for how this can be done. The United Kingdom's Human Fertilisation and Embryology Authority is probably the world's best example of a government regulatory body thoughtfully overseeing advanced reproductive technologies. Different countries can have different models, but every country should have something. When they do, they'll be better able to share engagement and regulatory models, learn from each other, and together evolve best practices. These UK government efforts have been supplemented by the excellent work of the private charities Private Education Trust, Genetic Alliance UK, and the Wellcome Trust, engaging the British public and developing language supporting a more inclusive public dialogue.[21] The United Kingdom's Nuffield Council on Bioethics in 2018 recommended the creation of an independent government-funded body to promote public understanding and engagement around issues of genome editing and human reproduction.[22] Denmark also does a great job empowering citizen groups to provide input to elected representatives about genetic engineering and other technological issues.

But even the most internationally minded people among us cannot and should not wait for our countries and the international community to act. Each of us must take individual responsibility for ensuring we are educating ourselves about what's coming and bringing as many of our fellow humans as possible into the conversation. We must each play a role in jump-starting a species-wide conversation on the future of human genetic engineering before it's too late.

Long before the international community turned against slavery, land

---

* Harvard biologist and naturalist E. O. Wilson made a similar point when he wrote, "We have created a Star Wars civilization, with Stone Age emotions, medieval institutions, and godlike technology." Futurist visionary Stewart Brand makes a similar point: "We are as gods and might as well get good at it."

mines, and blood diamonds, or embraced anti-corruption, African debt
relief, and environmental protection, activists lit a spark that inspired
popular movements, which, in turn, pressured governments in each of
these areas.

Norms are, in many ways, squishy and difficult to measure. Slavery, for
example, was widely accepted in many parts of the world until the antislav-
ery movement caught steam in nineteenth-century England. From a small
spark at Cambridge University, the movement grew until British laws were
changed, slavery was abolished in the United States and elsewhere, and
the idea of slavery became increasingly taboo. Before the idea of antislav-
ery was translated into law, it was a general feeling, a growing zeitgeist.
Lots of individual conversations, deep thinking, research, soul searching,
dialogues, coalition building, and ultimately fights and even civil wars gave
the idea form, like a rolling snowball.

"Never doubt," the cultural anthropologist Margaret Mead (allegedly)
once said, "that a small group of thoughtful, committed citizens can change
the world. Indeed, that is the only thing that ever has." This type of norm-
building process is required to help our species articulate and champion
our best values to find the right balance on human genetic engineering. We
must foster a continual series of local, national, and global conversations
that can, over time, congeal into global norms and lay an informed foundation for the decisions we will collectively need to make in the future.

We've never conducted a species-wide conversation on our future. But
how could we have organized a global dialogue in the eighteenth century
at the advent of the industrial revolution when only around 10 percent of
the global population was literate and the primary means of international
communication was snail mail that took months to deliver? Even in 1945,
at the dawn of the nuclear age, only around 30 percent of the world population was literate and landline telephones were few and far between. Today,
about 85 percent of the global population is literate, two-thirds have
their own cell phones, and over half have access to the internet—and the
numbers are growing.

With an increasing percentage of the world population connected to the information grid in one way or another, we now have an unprecedented opportunity to come together for a more meaningful collective process than ever before. Given the trajectory of the science and political divisiveness of the issues, the window for constructive dialogue may not stay open long.

A species-wide conversation would involve connecting individuals and communities around the world with different backgrounds and perspectives and varying degrees of education in an interconnected web of dialogue. It would link people adamantly opposed to human genetic enhancement, those who may see it as a panacea, and the vast majority of everyone else who has no idea this transformation is already underway. It would highlight the almost unimaginable positive potential of these technologies but also be honest and straightforward about the potential dangers.

One way of structuring this global dialogue would be to establish an international commission of top scientists, thinkers, religious leaders, and others tasked to come up with a discrete number of essential questions about the future of human genetic engineering. These questions might include:

1. What can be done to ensure the broadest possible access to the health and well-being benefits of genetic technologies?
2. Should there be limits on the application of genetic technologies to treat or eliminate disease? If so, what should they be?
3. Should people have full access to information about their own and their potential future children's genetic makeup or should this access be limited? If limited, what should the restrictions be and why?
4. Should parents be allowed unlimited freedom to select from among their natural embryos during in vitro fertilization? If not, on what basis should limitations be set? Should parents be allowed to select embryos based on non-disease-related traits, like height, projected IQ, personality style, etc.?
5. If it is proven safe, should precision gene editing be used to eliminate

genetic diseases in adult sex cells and preimplanted embryos in a manner that would be passed to future generations?

6. Do we need a global framework to help prevent the worst abuses of human genetic engineering? If so, what standards should underpin these efforts?

7. What long-term institutions do we need to foster an inclusive global dialogue on the future of human genetic engineering that optimizes the benefits of these technologies and minimizes potential harms?

8. What more can be done to help ensure the genetic revolution helps enhance all of our humanity, and how can each of us be better engaged in this process?

After these issues are framed by the expert commission, easily understood multimedia background materials could be created for each question that could be a foundation for an extensive and ongoing series of dialogues to be held around the world.

A global coalition of partner universities, school systems, think tanks, religious organizations, and civil society groups could then organize these dialogues to be held in venues big and small all around the world. Each partner organization could organize its own forums as well as real and virtual dialogues based on the questions and report back their findings to the commission.

On an even more democratic level, motivated individuals and any other type of group could be encouraged to self-organize their own conversations based on the same central questions. These hundreds of thousands or even millions of conversations could be held in conference rooms, classrooms, place of worship, kitchens, bus stops, barber shops, town squares, nail salons, union halls, internet chat rooms, village commons, virtual worlds, social media forums, and countless other places where humans connect.

Global virtual summits could periodically be held to bring together populations from all of the various dialogues around the world, and a network of talks and content based on the TEDx model could both help make learning and participating interesting, exciting, and accessible, and raise additional core questions needing to be addressed.

When this structured and informed global dialogue on the future of human genetic engineering reaches critical mass, an ongoing dialogue mechanism—possibly a new body connected to the United Nations—could be created to help further systematize the process and take it to a next level. This organization could pull together the content from the global conversations, views of experts, and inputs from national, nongovernmental, and other communities into an ongoing series of recommendations constantly churning through the cycle of public and expert engagement.

Over time, this type of engaged process could help individuals, societies, states, and the global community writ large better understand the genetic and other scientific revolutions, become more active participants in the decision-making process about our common future, and begin to define limited redlines beyond which we collectively feel our species should not, for the moment at least, go.

A species-wide dialogue on the future of human genetic engineering may seem like a drop in the bucket compared to the magnitude of the challenge the genetics revolution will bring. Such a process could even potentially do harm by waking up slumbering neo-Luddites, who might forcibly oppose even the most benign and beneficial genetic technologies. But the alternative to getting started with this kind of broad, public engagement is worse. If a relatively small number of even well-intentioned specialists unleash a human genetic revolution that touches most everyone else and ultimately alters our species' evolutionary trajectory without informed, meaningful, and early input from others, the backlash against the genetic revolution will overwhelm its monumental potential for good. Humans must and will embrace the genetic revolution, but we will be far better off if we do it together.

※

The genetics revolution will unlock one of the greatest opportunities for advancing human health and well-being in the history of our species. We will demand access to genetic technologies for ourselves and our children as a next step in our perpetual struggle with the cruelty of the natural world and to realize our greatest aspirations to transcend our limited biology and, someday, our time-limited planet.

Figuring out how to deploy genetic technologies in ways that enhance our dignity and respect for each other will require us to draw on the best of our humanist values and double down on our embrace of, respect for, and investment in our diversity, equality, and common humanity. While the genetic engineering technologies are new, the values and philosophies we will need to use them wisely are often very old.

Deploying our best values at this transitional moment for our species demands that we all understand what is happening now, what is coming, what's at stake, and the role we each must play in building a technologically enhanced future that works for all of us.

Jump-starting dialogues to develop norms that translate into international best practices and, ultimately, global regulations will be long and arduous. It all might end up being impossible. But simply trying will bring more people into the processes of determining the future of humanity. We will not be able to stop the genetic enhancement of our species, but we can influence, hopefully for the better, how this transformation plays out.

It will be a difficult, painful, and conflict-ridden process, but we have no alternative. We all need to participate. We don't have a moment to lose in getting started.

Now that you have read this book, you are a critical catalyst of this dialogue.

Fellow humans, let us together begin the conversation.

Notes

INTRODUCTION

1   Harvard biologist Michael Desai's fascinating research on yeast that can reproduce both clonally and sexually has shown how sexually reproducing yeast adapt two times faster than their asexually reproducing landsmen, allowing harmful mutations to be reduced and helpful ones spread. Michael J. McDonald, Daniel P. Rice, and Michael M. Desai. "Sex Speeds Adaptation by Altering the Dynamics of Molecular Evolution." *Nature* 531, no. 7593 (2016): 233–36. doi:10.1038/nature17143.

2   "Genetics and Other Human Modification Technologies: Sensible International Regulation or a New Kind of Arms Race?" Hearing before the Subcommittee on Terrorism, Nonproliferation, and Trade of the Committee on Foreign Affairs, House of Representatives, One Hundred Tenth Congress, Second Session, June 19, 2008, Serial No. 110–201. https://fas.org/irp/congress/2008_hr/genetics.pdf.

CHAPTER 1

1   For an incredible visual image of this, see Courtney K. Ellison, Triana N. Dalia, Alfredo Vidal Ceballos, Joseph Che-Yen Wang, Nicolas Biais, Yves V. Brun, and Ankur B. Dalia. "Retraction of DNA-Bound Type IV Competence Pili Initiates DNA Uptake during Natural Transformation in Vibrio Cholerae." *Nature Microbiology*, 2018, doi:10.1038/s41564-018-0174-y.

2   Priya Verma, "Reproduction and the Discovery of Sperm," Elawoman, accessed June 25, 2018, https://elawomn.quora.com/Reproduction-and-the-Discovery-of-Sperm.

3   F. Gilbert, "Structure of the Gametes," in *Developmental Biology*, 6th edition. (Sunderland, MA: Sinauer Associates, 2000), https://www.ncbi.nlm.nih.gov/books/NBK10005/.

4   A few anthropologists make the case that humans began domesticating crops around 23,000 years ago. See Ainit Snir, Dani Nadel, Iris Groman-Yaroslavski, Yoel Melamed, Marcelo Sternberg, Ofer Bar-Yosef, and Ehud Weiss. "The Origin of Cultivation and Proto-Weeds, Long Before Neolithic Farming," *Plos One* 10, no. 7 (2015), doi:10.1371/journal.pone.0131422.

5   Most of these epigenetic marks develops from scratch in a newly conceived organism, but recent research has shown that a small percentage of them are actually inherited from an organism's parents. This has led some people to question the prior belief that the theories of Jean-Baptiste de Lamarck were entirely wrong.

6   For a more technical history of DNA sequencing, see James M. Heather and Benjamin Chain, "The Sequence of Sequencers: The History of Sequencing DNA," *Genomics* 107, no. 1 (2016): 1–8, doi:10.1016/j.ygeno.2015.11.003.

7   John J. Kasianowicz and Sergey M. Bezrukov, "On Three Decades of Nanopore Sequencing," *Nature Biotechnology* 34 (2016): 481–482.

8   "OMIM Entry Statistics," last modified July 24, 2018, http://www.omim.org/statistics/entry.

9   See Jason L. Vassy et al., "The Impact of Whole-Genome Sequencing on the Primary Care and Outcomes of Healthy Adult Patients," *Annals of Internal Medicine* 167, no. 3 (2017): 159, doi:10.7326/m17–0188; and P. Natarajan et al., "Aggregate Penetrance of Genomic Variants for Actionable Disorders in European and African Americans," *Science Translational Medicine* 8, no. 364 (2016), doi:10.1126/scitranslmed.aag2367.

10   Rachel D. Melamed et al., "Genetic Similarity between Cancers and Comorbid Mendelian Diseases Identifies Candidate Driver Genes," *Nature Communications*, no. 1 (2015), doi:10.1038/ncomms8033.

11   See Lidewij Henneman et al., "Responsible Implementation of Expanded Carrier Screening," *European Journal of Human Genetics* 24, no. 6 (June 2016): e1–e12. Published online 2016 Mar 16. doi: 10.1038/ejhg.2015.271; and Nancy C. Rose, "Expanded Carrier Screening: Too Much of a Good Thing?" *Prenatal Diagnosis* 35, no. 10 (2015): 936–37, doi:10.1002/pd.4638.

12   Heather Mason Kiefer, "Gallup Brain: The Birth of In Vitro Fertilization," *Gallup News*, August 5, 2003, http://news.gallup.com/poll/8983/gallup-brain-birth-vitro-fertilization.aspx.

13   Kiefer, "Gallup Brain."

14   John M. Haas, "Begotten Not Made: A Catholic View of Reproductive Technology," United States Conference of Catholic Bishops, 1998, http://www.usccb.org/issues-and-action/human-life-and-dignity/reproductive-technology/begotten-not-made-a-catholic-view-of-reproductive-technology.cfm.

15   Jason Pontin, "Science Is Getting Us Closer to the End of Infertility," *Wired*, March 27, 2018, https://www.wired.com/story/reverse-infertility/.

16   G. Edwards and R. L. Gardner, "Sexing of Live Rabbit Blastocysts," *Nature* 214 (1967): 576–77, https://www.nature.com/articles/214576a0.

17   Genome Editing and Human Reproduction. Report. July 17, 2018, Nuffield Council on Bioethics. http://nuffieldbioethics.org/wp-content/uploads/Genome-editing-and-human-reproduction-FINAL-website.pdf, p. 10.

18   Henry T. Greely, *The End of Sex and the Future of Human Reproduction* (Cambridge: Harvard University Press, 2016): 114, 331.

19   The mathematics are a bit more complicated than this preliminary calculation suggests because PGT is more accurate with some categories of genetic disorders than others. Recessive conditions where both parents are heterozygous for the particular disorder, for

example, would be statistically easier to identify than a dominant condition where one parent is a heterozygous carrier and the other is a noncarrier.

20   Petula Dvorak, "Parents Who Refuse to Vaccinate Their Children Are Putting Others at Risk," *Washington Post,* January 26, 2015, https://www.washingtonpost.com/local/parents -who-refuse-to-vaccinate-their-children-are-putting-others-at-risk/2015/01/26/9c538266 -a5aa-11e4-a06b-9df2002b86a0_story.html?utm_term=.3b17e2cf1ccf.

21   "Vaccines Do Not Cause Autism," Centers for Disease Control, last modified November 23, 2015, https://www.cdc.gov/vaccinesafety/concerns/autism.html.

22   Laura Entis, "Donald Trump Has Long Linked Autism to Vaccines. He Isn't Stopping Now That He's President," *Fortune,* February 16, 2017, http://fortune.com/2017/02/16 /donald-trump-autism-vaccines/.

23   Cary Funk, Brian Kennedy, and Meg Hefferon, "Public Opinion about Childhood Vaccines for Measles, Mumps, and Rubella," Pew Research Center, February 2, 2017, http://www.pewinternet.org/2017/02/02/public-opinion-about-childhood-vaccines -for-measles-mumps-and-rubella/.

24   Norbert Gleicher and Raoul Orvieto, "Is the Hypothesis of Preimplantation Genetic Screening (PGS) Still Supportable? A Review," *Journal of Ovarian Research* 10 (2017), https://www.ncbi.nlm.nih.gov/pmc/articles/PMC5368937/. A major review of 455 publications on genetic screening, published in March 2017, asserted there was a higher than appreciated rate of false positives when screening early-stage human embryos for genetic abnormalities, because both cells with these mutations were spread unevenly across the embryos and because doctors didn't fully appreciate the ability of the embryos to self-correct.

25   Julian Quinones and Arijeta Lajka, "'What Kind of Society Do You Want to Live In?': Inside the Country Where Down Syndrome Is Disappearing," *CBS News,* August 15, 2017, https://www.cbsnews.com/news/down-syndrome-iceland/.

26   Susannah Maxwell, Carol Bower, and Peter O'Leary, "Impact of Prenatal Screening and Diagnostic Testing on Trends in Down Syndrome Births and Terminations in Western Australia 1980 to 2013," *Prenatal Diagnosis* 35 (2015): 1324–1330; Manya Koetse, "The Last Downer: China and the End of Down Syndrome," *What's On Weibo,* May 2, 2016, https://www.whatsonweibo.com/china-end-syndrome/; Polina Bachlakova, "Why Are 95% of Danish Women Aborting Babies with Severe Developmental Disabilities?" *VICE News,* April 24, 2015, https://www.vice.com/da/article/ex9daw/why-are-95-of-danish -women-aborting-babies-with-severe-developmental-disabilities-020; Myrthe Jacobs et al., "Pregnancy Outcome Following Prenatal Diagnosis of Chromosomal Anomaly: A Record Linkage Study of 26,261 Pregnancies," *Plos One* 11 (2016).

27   Yuval Levin, "Public Opinion and the Embryo Debates," *The New Atlantis* 20 (2008): 47– 62, https://www.thenewatlantis.com/publications/public-opinion-and-the-embryo-debates.

28   Jaime L. Natoli et al., "Prenatal Diagnosis of Down Syndrome: A Systematic Review of Termination Rates (1995–2011)," *Prenatal Diagnosis* 32 (2012): 142–153, http:// onlinelibrary.wiley.com/doi/10.1002/pd.2910/abstract;jsessionid=48749E1687B81 B0E20AD95934451778E.f01t03.

29   Henry T. Greely, *The End of Sex and the Future of Human Reproduction* (Cambridge: Harvard University Press, 2016).

30   Kees van Gool et al., "Understanding the Costs of Care for Cystic Fibrosis: An Analysis by Age and Health State," *Value in Health* 16 (2013): 345–355, https://www.sciencedirect .com/science/article/pii/S1098301512042684. A 2010 study was far more aggressive in its estimates of the cost savings that would come from using PGT to prevent cystic fibrosis. Claiming that the average additional cost of care for a person with CF was $63,127 for an average life expectancy of thirty-seven years, the authors estimate the cumulative savings over thirty-seven years as $33.3 billion, a far larger number than mine. Additionally, according to the authors, "a total of 618,714 cumulative years of patients suffering because of CF and thousands of abortions could be prevented." These higher estimates strengthen the cost-benefit case for universal embryo screening. I. Tur-Kaspa et al., "PGT for All Cystic Fibrosis Carrier Couples: Novel Strategy for Preventive Medicine and Cost Analysis." *Reproductive BioMedicine Online* 21, no. 2 (2010): 186–95. doi:10.1016/j.rbmo.2010.04.031.

31   Victoria Divino et al., "The Direct Medical Costs of Huntington's Disease by Stage: A Retrospective Commercial and Medicaid Claims Data Analysis," *Journal of Medical Economics* 16 (2013): 1043–1050, https://www.researchgate.net/publication/241690 299_The_Direct_Medical_Costs_of_Huntington's_Disease_by_Stage_A_Retrospective _Commercial_and_Medicaid_Claims_Data_Analysis.

32   Using CDC estimates, David Sable calculates that 150,000 affected children are born annually, of which 30,000 are congenital. See David Sable, "Why the Future of Precision Medicine Runs through the IVF Lab." *Forbes*, April 22, 2018. https://bit.ly/2FF2NaS.

33   "Findings Suggest Increased Number of IVF Cycles Can Be Beneficial," *The JAMA Network,* December 22, 2015, https://media.jamanetwork.com/news-item/findings -suggest-increased-number-of-ivf-cycles-can-be-beneficial/.

34   Said M. Yildiz and M. Mahmud Khan, "Opportunities for Reproductive Tourism: Cost and Quality Advantages of Turkey in the Provision of In-Vitro Fertilization (IVF) Services," *BMC Health Services Research* 16 (2016): 378, https://www.ncbi.nlm.nih.gov /pmc/articles/PMC4982316/.

35   Ido Efrati, "Israel Remains an IVF Paradise as Number of Treatments Rises 11% in 2016," *Haaretz*, May 11, 2017, https://www.haaretz.com/israel-news/.premium-israel-remains -as-ivf-paradise-as-number-of-treatments-rises-1.5470164.

36   David Sable, "The Seven Trends That Define the Future of IVF," *Forbes,* February 28, 2015, https://www.forbes.com/sites/davidsable/2015/02/28/the-seven-trends-that -define-the-future-of-ivf/#67c863c8494b.

37   Claire Cain Miller, "Freezing Eggs as Part of Employee Benefits: Some Women See Darker Message," *New York Times*, October 14, 2014, https://www.nytimes.com/2014/10/15 /upshot/egg-freezing-as-a-work-benefit-some-women-see-darker-message.html.

38   Charlotte Alter, "Sheryl Sandberg Explains Why Facebook Covers Egg-Freezing," *Time*, April 25, 2015, http://time.com/3835233/sheryl-sandberg-explains-why-facebook -covers-egg-freezing/.

39   The FertilityIQ Family Builder Workplace Index: 2017–2018, https://www.fertilityiq .com/fertilityiq-data-and-notes/fertilityiq-best-companies-to-work-for-family-builder -workplace-index-2017–2018.

40   I recognize that the widespread adoption of preimplantation embryo screening

to date has been less than might have been predicted. This is likely because the procedure is in its early stage of development, expensive, and not covered by insurance in many jurisdictions, and because medical and social norms are still evolving. It may be that some jurisdictions adopt PGT at different rates, but it is my contention that adoption rates will increase over time everywhere, with the biggest jumps in places like China.

CHAPTER 2

1   For more on "multi-omics" systems biology, see Yehudit Hasin, Marcus Seldin, and Aldons Lusism, "Multi-Omics Approaches to Disease." *Genome Biology*, May 5, 2017, accessed June 26, 2018, https://genomebiology.biomedcentral.com/articles/10.1186 /s13059-017-1215-1, and Marc Santolini et al., "A Personalized, Multiomics Approach Identifies Genes Involved in Cardiac Hypertrophy and Heart Failure." *NPJ Systems Biology and Applications* 4, no. 1 (2018), doi:10.1038/s41540-018-0046-3.

2   Erika Check Hayden, "Human Genome at Ten: Life Is Complicated," *Nature* 464 (2010): 664–667, https://www.nature.com/news/2010/100331/full/464664a.html.

3   Hayden, "Human Genome at Ten."

4   Holley, Peter, "Elon Musk's Nightmarish Warning: AI Could Become 'An Immortal Dictator from Which We Would Never Escape,'" *Washington Post*, April 6, 2018, https://www.washingtonpost.com/news/innovations/wp/2018/04/06/elon-musks -nightmarish-warning-ai-could-become-an-immortal-dictator-from-which-we-would -never-escape/?noredirect=on&utm_term=.b662dac58897.

5   "How Much Data Does the World Generate Every Minute?" IFL Science, http://www .iflscience.com/technology/how-much-data-does-the-world-generate-every-minute/.

6   Deep Genomics, https://www.deepgenomics.com/.

7   Sara Castellanos, "Quantum Computing May Speed Drug Discovery, Biogen Test Suggests," *Wall Street Journal*, June 13, 2017, https://blogs.wsj.com/cio/2017/06/13 /quantum-computing-may-speed-drug-discovery-biogen-test-suggests/.

8   "The World's Most Valuable Resource Is No Longer Oil, but Data," *The Economist*, May 6, 2017, https://www.economist.com/news/leaders/21721656-data-economy-demands -new-approach-antitrust-rules-worlds-most-valuable-resource.

9   Stephen Hsu, "Genomic Prediction of Complex Traits", lecture, The Paul G. Allen Frontiers Group, January 19, 2018, http://infoproc.blogspot.com/2018/01/allen -institute-meeting-on-genetics-of.html.

10 Michael D. Lemonick, "The Iceland Experiment," *Time*, February 12, 2006, http:// content.time.com/time/printout/0,8816,1158968,00.html#.

11 "Harnessing the Power of Genomics through Global Collaborations and Scientific Innovation," AstraZeneca, January 12, 2018, accessed August 5, 2018, https://www .astrazeneca.com/media-centre/articles/2017/harnessing-the-power-of-genomics -through-global-collaborations-and-scientific-innovation-12012018.html.

12 "The 1,000 Genomes Project," Genomics England, https://www.genomicsengland .co.uk/the-100000-genomes-project/.

13 All of Us, National Institutes of Health, https://allofus.nih.gov/; Gina Kolata, "The

Struggle to Build a Massive 'Biobank' of Patient Data," *New York Times,* March 19, 2018, https://www.nytimes.com/2018/03/19/health/nih-biobank-genes.html; Megan Molteni, "The NIH Launches Its Ambitious Million-Person Genetic Survey," *Wired,* May 5, 2018, accessed June 1, 2018, https://www.wired.com/story/all-of-us-launches/.

14   As of 2018, around 600,000 veterans had been sequenced. "Million Veteran Program (MVP)," Office of Research and Development, U.S. Department of Veterans Affairs, https://www.research.va.gov/mvp/.

15   Luna DNA, accessed June 17, 2018, https://www.lunadna.com/. Nebula Genomics also uses blockchain technology to help consumers understand how their genetic information is being used and benefit from commercial activity surrounding that information. "Nebula Genomics," accessed August 27, 2018, https://www.nebulagenomics.io/.

16   Founded in 2005, the Personal Genome Project is an international coalition of genome projects using an open-source, volunteer strategy to bring together vast pools of genetic data matched with medical and other life records. As of March 2018, partners in the United States, Canada, United Kingdom, Austria, and China have together pooled data on a mere 522 genomes, according to Madeleine Price Ball, cofounder of the Open Humans Project, which runs the Personal Genomes Project.

17   David Cyranoski, "China's Bid to Be a DNA Superpower," *Nature* 534 (2016): 462–463, http://www.nature.com/news/china-s-bid-to-be-a-dna-superpower-1.20121.

18   China's Jiangsu province, to give one example, is partnering with the National Health and Family Planning Commission to sequence the genes of one million human subjects and create the largest sequencing platform and biomedical big data analysis center in Asia. Chinese companies BGI Shenzhen and WuXi NextCode, an offshoot of Iceland's deCODE, are also building massive genomic databases. See Dou Shicong, "Jiangsu Government Unveils Plans to Sequence Genes of 1 Million Subjects," *Yicai Global,* October 30, 2017, https://www.yicaiglobal.com/news/jiangsu-government -unveils-plans-sequence-genes-1-million-subjects; "Jiangsu Officially Launches Million Population Genome Sequencing Project," *China News,* October 29, 2017, http://www .chinanews.com/gn/2017/10–29/8362987.shtml; Preetika Rana, "Made-to-Order Medicine: China, U.S. Race to Decode Your Genes," *Wall Street Journal,* September 20, 2017, https://www.wsj.com/articles/china-rushes-to-surpass-u-s-in-decoding-citizens -genes-1505899806; David Cyranoski, "China Embraces Precision Medicine on a Massive Scale," *Nature News,* January 6, 2016, https://www.nature.com/news/china -embraces-precision-medicine-on-a-massive-scale-1.19108.

19   Bertil Schmidt and Andreas Hildebrandt, "Next-Generation Sequencing: Big Data Meets High Performance Computing," *Drug Discovery Today* 22 (2017): 712–717.

20   Valeria D'Argenio, "The High-Throughput Analyses Era: Are We Ready for the Data Struggle?" *High-Throughput* 7 (2018): 8.

21   "White House Precision Medicine Initiative," National Archives and Records Administration, https://obamawhitehouse.archives.gov/node/333101. The Precision Medicine Initiative later morphed into the Cancer Moonshot and All of Us initiatives. William Lane and his colleagues released an important study in 2018 showing how whole genome sequencing could do a far more precise job blood typing for transfusions

than the traditional blood-typing model, potentially saving many lives. Robert Flower, Eileen Roulis, and Catherine Hyland, "Whole-Genome Sequencing Algorithm for Blood-Group Typing," *The Lancet Haematology* 5, no. 6 (2018): doi:10.1016/s2352 –3026(18)30064–4.

22 Ellie Kincaid, "Geisinger Says DNA Sequencing as Preventative Care Is Ready for the Clinic," *Forbes*, May 9, 2018, accessed May 12, 2018, https://www.forbes.com/sites /elliekincaid/2018/05/07/geisinger-says-dna-sequencing-as-preventative-care-is-ready -for-the-clinic/#60fc3ba34d63.

23 Robert Green et al., "Whole-Genome Sequencing in Primary Care," *Annals of Internal Medicine* 167, no. 3 (2017): doi:10.7326/p17–9040.

24 Amit V. Khera et al., "Genome-Wide Polygenic Scores for Common Diseases Identify Individuals with Risk Equivalent to Monogenic Mutations," *Nature Genetics* 50 (August 13, 2018): doi:10.1038/s41588-018-0183-z.

25 "About Genomic Prediction," Genomic Prediction, accessed August 5, 2018, https:// genomicprediction.com/about/.

CHAPTER 3

1 Bouchard et al., "Sources of Human Psychological Differences: The Minnesota Study of Twins Reared Apart," *Science* 250 (1990): 223–228, https://www.ncbi.nlm.nih.gov /pubmed/2218526.

2 Bouchard et al., "Sources of Human Psychological Differences."

3 J. Polderman et al., "Meta-Analysis of the Heritability of Human Traits Based on Fifty Years of Twin Studies," *Nature Genetics* 47 (2015): 702–709, https://www.ncbi.nlm.nih .gov/pubmed/25985137.

4 Daniel Schwekendiek, "Determinants of Well-Being in North Korea: Evidence from the Post-Famine Period," *Economics & Human Biology* 6 (2008): 446–454. The malnourishment of a pregnant mother can lead to lower cognitive function, shorter stature, immune deficiencies, shorter life spans, and many other problems. Caroline H. D. Fall, "Fetal Malnutrition and Long-Term Outcomes," in *Maternal and Child Nutrition: The First 1,000 Days*, 74th Nestlé Nutrition Institute Workshop Series (2013): 11–25, doi:10.1159/000348384.

5 Chao-Qiang Lai, "How Much of Human Height Is Genetic and How Much Is Due to Nutrition?" *Scientific American*, December 11, 2006, https://www.scientificamerican .com/article/how-much-of-human-height/.

6 Louis Lello et al., "Accurate Genomic Prediction of Human Height," *BioRxiv* (2017), https://www.biorxiv.org/content/early/2017/10/07/190124.

7 Lello et al.; Stephen Hsu, "Genomic Prediction of Complex Traits," lecture, The Paul G. Allen Frontiers Group, January 19, 2018, http://infoproc.blogspot.com/2018/01/allen -institute-meeting-on-genetics-of.html.

8 Ian J. Deary, "Intelligence," *Current Biology* 23 (2013): R673–R676, http://www .cell.com/current-biology/fulltext/S0960–9822(13)00844–0; Yan Zhang et al., "Estimation of Complex Effect-Size Distributions Using Summary-Level Statistics from

Genome-wide Association Studies across 32 Complex Traits and Implications for the Future," *BioRxiv* (2017), https://www.biorxiv.org/content/biorxiv/early/2017/08/11/175406.full.pdf; Stephen Hsu, "Genomic Prediction of Complex Traits."

9   Yan Zhang et al., "Estimation of Complex Effect-Size Distribution."

10   Deary, "Intelligence."

11   Charles Spearman, "'General Intelligence,' Objectively Determined and Measured," *American Journal of Psychology* 15, no. 2 (1904): 201–93, http://dx.doi.org/10.2307/1412107.

12   Brian Resnick, "IQ, Explained in 9 Charts," *Vox*, October 10, 2017, https://www.vox.com/2016/5/24/11723182/iq-test-intelligence. Robert Plomin and Sophie Von Stumm, "The New Genetics of Intelligence," *Nature Reviews Genetics* 19, no. 3 (2018): 148–59, doi:10.1038/nrg.2017.104.

13   Patrick F. McKay et al., "The Effects of Demographic Variables and Stereotype Threat on Black/White Differences in Cognitive Ability Test Performance," *Journal of Business and Psychology* 18 (2003): 1–14, https://link.springer.com/article/10.1023/A:1025062703113. See also Daphne Martschenko, "The IQ Test Wars: Why Screening for Intelligence Is Still So Controversial," *The Conversation*, October 10, 2017, accessed June 26, 2018, https://theconversation.com/the-iq-test-wars-why-screening-for-intelligence-is-still-so-controversial-81428.

14   Richard J. Herrnstein and Charles A. Murray, *The Bell Curve: Intelligence and Class Structure in American Life* (New York: Free Press, 1997). This book was first published in 1994.

15   Bob Herbert, "In America; Throwing a Curve," *New York Times*, October 26, 1994, accessed June 26, 2018. https://www.nytimes.com/1994/10/26/opinion/in-america-throwing-a-curve.html.

16   Stephen Jay Gould, "Curveball," *New Yorker*, November 28, 1994, http://www.dartmouth.edu/~chance/course/topics/curveball.html. For a thoughtful, strong critique of *The Bell Curve* see also Siddhartha Mukherjee, *The Gene: An Intimate History* (London: Vintage, 2017), 343.

17   For a compilation of some of these critiques, see Steve Fraser, *The Bell Curve Wars: Race, Intelligence, and the Future of America* (New York: BasicBooks, 1998). See also Eric Siegel, "The Real Problem with Charles Murray and 'The Bell Curve,'" Scientific American Blog Network, April 12, 2017, accessed June 17, 2018. https://blogs.scientificamerican.com/voices/the-real-problem-with-charles-murray-and-the-bell-curve/.

18   Reprinted in Linda S. Gottfredson, "Mainstream Science on Intelligence: An Editorial with 52 Signatories, History, and Bibliography," *Intelligence* 24, no. 1 (1997): 13–23, doi:10.1016/s0160–2896(97)90011–8. The American Psychological Association also created a special task force to assess the findings of the book that arrived at similar results.

19   Bouchard et al., "Sources of Human Psychological Differences."

20   Plomin and I. J. Deary, "Genetics and Intelligence Differences: Five Special Findings," *Molecular Psychiatry* 20 (2014): 98–108, https://www.nature.com/articles/mp2014105.pdf.

21   Thomas J. Bouchard, "The Wilson Effect: The Increase in Heritability of IQ with Age," *Twin Research and Human Genetics* 16, no. 5 (2013): 923–30, doi:10.1017/thg.2013.54. Because IQ is a relative term, genetics would need to be a more important factor in IQ as we age relative to other people as they age. Another study suggested that adult IQ is 80

percent heritable. Valerie S. Knopik et al., *Behavioral Genetics*, 7th ed. (Worth Publishers, 2016); Susan Bouregy, Elena L. Grigorenko, and Stephen R. Latham, *Genetics, Ethics and Education* (Cambridge: Cambridge University Press, 2017).

22   Eric Turkheimer et al., "Socioeconomic Status Modifies Heritability of IQ in Young Children," *Psychological Science* 14 (2003): 623–628.

23   Mutations and variants are essentially the same thing, but we tend to call them *mutations* when they are disease-related and often rarer, and we call them *variants* when they impact nondisease traits and are more common. Some researchers have recently begun using the term *pathogenic variant* instead of *mutation* to make this point. Because the line between disorders and traits is so squishy, these terms are largely interchangeable.

24   D. Hill et al., "A Combined Analysis of Genetically Correlated Traits Identifies 187 Loci and a Role for Neurogenesis and Myelination in Intelligence," *Molecular Psychiatry* (2018): doi:10.1038/s41380-017-0001-5, https://www.nature.com/articles/s41380 -017-0001-5; D. Zabaneh et al., "A Genome-Wide Association Study for Extremely High Intelligence," *Molecular Psychiatry* 23 (July 4, 2017): 1226–1232, https://www .nature.com/articles/mp2017121; Gail Davies et al., "Ninety-Nine Independent Genetic Loci Influencing General Cognitive Function Include Genes Associated with Brain Health and Structure," *BioRxiv* (2017), https://www.biorxiv.org/content/biorxiv /early/2017/08/18/176511.full.pdf; Robert Plomin and Sophie von Stumm, "The New Genetics of Intelligence," *Nature Reviews Genetics* 19 (January 8, 2018): 148–159, doi:10.1038/nrg.2017.104, https://www.nature.com/articles/nrg.2017.104. The number will likely exceed two hundred by the time this book is published.

25   Suzanne Sniekers et al., "Genome-Wide Association Meta-analysis of 78,308 Individuals Identifies New Loci and Genes Influencing Human Intelligence," *Nature Genetics* 49 (2017): 1107–1112.

26   Already, researchers are moving from the bottom-up model of inferring general intelligence from identifying single-gene mutations to a more top-down approach, based on the big-data analytics of patterns identified in all of these mutations. See Plomin and Stumm, "The New Genetics of Intelligence."

27   Yan Zhang et al., "Estimation of Complex Effect-Size Distributions Using Summary-Level Statistics from Genome-wide Association Studies across 32 Complex Traits and Implications for the Future," *BioRxiv* (2017), https://www.biorxiv.org/content/biorxiv /early/2017/08/11/175406.full.pdf; Stephen Hsu, "Genomic Prediction of Complex Traits," lecture, The Paul G. Allen Frontiers Group, January 19, 2018, http://infoproc .blogspot.com/2018/01/allen-institute-meeting-on-genetics-of.html; "Heritability," *SNPedia*, last modified March 13, 2018, https://www.snpedia.com/index.php /Heritability. A preliminary list of select diseases and human body conditions estimated to be on average 50 percent or more genetic on SNPedia.com, a genetics wiki that brings together scientific research from around the world, includes: abdominal aortic aneurysm, acne, age-related macular degeneration, alcoholism, Alzheimer's disease, androgenic alopecia, anorexia, attention deficit hyperactivity disorder, beard thickness, bipolar disorder, bone mineral density, celiac disease, chronic obstructive pulmonary disease, Crohn's disease, Dupuytren's disease, eczema, epilepsy, eye color, freckle counts, Graves' disease, hair color, hair curliness, height, kidney stones, lupus, menarche, monobrow,

polycystic ovary syndrome, psoriasis, rheumatoid arthritis, schizophrenia, sexual orientation, stuttering, thyroid cancer, Tourette's syndrome, type 1 diabetes, and varicose veins. The full list will eventually include many tens of thousands of diseases and traits.

28 Bouchard et al., "Sources of Human Psychological Differences."

29 Min-Tzu Lo et al., "Genome-Wide Analyses for Personality Traits Identify Six Genomic Loci and Show Correlations with Psychiatric Disorders," *Nature Genetics* 49 (2017): 152–156, https://www.nature.com/articles/ng.3736.

CHAPTER 4

1 "Masturbatorium," *Wiktionary*, last modified June 7, 2017, https://en.wiktionary.org/wiki/masturbatorium.

2 "Odontophilia," Urban Dictionary, accessed June 26, 2018, https://www.urbandictionary.com/define.php?term=Odontophilia.

3 Japan has both the highest percentage of IVF births and the lowest success rate among industrialized countries. This is in part because the women receiving IVF tend to be older than their counterparts in other countries and because the IVF industry in Japan has inconsistent standards. See "A Corked Tube: No Country Resorts to IVF More Than Japan—or Has Less Success," *The Economist*, May 26, 2018, https://www.economist.com/asia/2018/05/26/no-country-resorts-to-ivf-more-than-japan-or-has-less-success.

4 Gary J. Gates, "In U.S., More Adults Identifying as LGBT," *Gallup*, January 11, 2017, http://news.gallup.com/poll/201731/lgbt-identification-rises.aspx. Measures of the exact size of the U.S. and global LGBTQIAP populations are inherently imperfect and do not account for the gender and sexual orientation fluidity, particularly common among younger generations. See: Katy Steinmetz, "Inside the Efforts to Finally Identify the Size of the Nation's LGBT Population," *Time*, May 18, 2016, http://time.com/lgbt-stats/.

5 Elizabeth Cohen, "Researchers Isolate Human Stem Cells in the Lab," CNN Interactive, November 5, 1998, http://www.cnn.com/HEALTH/9811/05/stem.cell.discovery/.

6 Jessica Reaves, "The Great Debate over Stem Cell Research," *Time*, July 11, 2001, http://content.time.com/time/nation/article/0,8599,167245,00.html.

7 Hayashi, "Offspring from Oocytes Derived from In Vitro Primordial Germ Cell-Like Cells in Mice," *Science* 16 (2012): 971–975, https://www.ncbi.nlm.nih.gov/pubmed/23042295. Interestingly, Hayashi is also leading efforts to use induced stem-cell technology to save the endangered white rhino from extinction.

8 David Cyranoski, "Rudimentary Egg and Sperm Cells Made from Stem Cells," *Nature*, December 24, 2014, https://www.nature.com/news/rudimentary-egg-and-sperm-cells-made-from-stem-cells-1.16636.

9 Naoko Irie, Shinseog Kim, and M. Azim Surani, "Human Germline Development from Pluripotent Stem Cells In Vitro," *Journal of Mammalian Ova Research* 33 (2016): 79–87; Naoko Irie and M. Azim Surani, "Efficient Induction and Isolation of Human Primordial Germ Cell-like Cells from Competent Human Pluripotent Stem Cells" in *Germline Stem Cells*, ed. Steven X. Hou and Shree Ram Singh, in *Methods in Molecular Biology* 1463 (2016): 217–226. Caution is also in order here. One of the four Yamanaka "master genes"—myc—is among the most aggressive catalysts of cancer.

10   *CIA World Factbook,* Central Intelligence Agency, accessed May 21, 2018. https://www
.cia.gov/library/publications/the-world-factbook/fields/2256.html.

11   Carl Shulman and Nick Bostrom, "Embryo Selection for Cognitive Enhancement:
Curiosity or Game-Changer?" *Global Policy* 5 (2014): 85–92, https://nickbostrom.com
/papers/embryo.pdf.

12   Stephen Hsu, "Super-Intelligent Humans Are Coming," *Nautilus,* October 16, 2014,
http://nautil.us/issue/18/genius/super_intelligent-humans-are-coming.

CHAPTER 5

1   "Method of the Year 2011," *Nature Methods* 9 (2012): 1, https://www.nature.com
/articles/nmeth.1852.

2   Eric S. Lander, "The Heroes of CRISPR," *Cell* 164 (2016): 18–28.

3   See "CRISPR Off-Targets: A Reassessment," *Nature News,* March 30, 2018, accessed May
8, 2018, https://www.nature.com/articles/nmeth.4664.

4   "SHERLOCK, DETECTR, CAMERA: Three New CRISPR Technologies.", AACC,
accessed   May   8,   2018,   https://www.aacc.org/publications/cln/cln-stat/2018
/march/15/sherlock-detectr-camera-three-new-crispr-technologies.

5   Pratiksha I. Thakore et al., "Editing the Epigenome: Technologies for Programmable
Transcription and Epigenetic Modulation," *Nature Methods* 13 (2016): 127–137, https://
www.ncbi.nlm.nih.gov/pmc/articles/PMC4922638/.

6   David Cano-Rodriguez and Marianne G. Rots, "Epigenetic Editing: On the Verge of
Reprogramming Gene Expression at Will," *Current Genetic Medicine Reports* 4 (2016):
170–179, https://link.springer.com/article/10.1007/s40142-016-0104-3; Heidi Ledford,
"CRISPR: Gene Editing Is Just the Beginning," *Nature* 531 (2016): 156–159, https://
www.nature.com/news/crispr-gene-editing-is-just-the-beginning-1.19510.

7   Emily Waltz, "Gene-Edited CRISPR Mushroom Escapes U.S. Regulation," *Nature* 532
(2016): 293, https://www.nature.com/news/gene-edited-crispr-mushroom-escapes-us
-regulation-1.19754.

8   Because neither the "Arctic apple" nor the nonbrowning mushrooms have foreign
DNA in them, they are technically not GMOs as traditionally defined. On March 28,
2018, U.S. Secretary of Agriculture Sonny Perdue announced that gene-edited plants
where no foreign DNA had been introduced would not be regulated as GMOs by
the USDA. This is causing consternation among many GMO advocates concerned
about any genetic alteration of the food supply. In November 2018, a coalition of
thirteen countries, including the United States, Canada, and Brazil, announced at a
World Trade Organization meeting in Geneva that they would support policies that
enable genome editing in agriculture. Amy Maxmen, "Genetically Modified Apple
Reaches U.S. Stores, but Will Consumers Bite?" *Nature* 551 (2017): 149–150, https://
www.nature.com/news/genetically-modified-apple-reaches-us-stores-but-will
-consumers-bite-1.22969. See also: "Secretary Perdue Issues USDA Statement on Plant
Breeding Innovation." USDA, March 28, 2018, accessed June 1, 2018, https://www
.usda.gov/media/press-releases/2018/03/28/secretary-perdue-issues-usda-statement
-plant-breeding-innovation and Emily Waltz, "CRISPR Crops—Exempt from GMO

Regulations—Reaching U.S. Market in Record Time," *Genetic Literacy Project,* January 15, 2018, https://geneticliteracyproject.org/2018/01/15/crispr-crops-exempt-gmo -regulations-reaching-us-market-record-time/. The European Court of Justice came to a similar conclusion in an opinion released in 2018: "Opinion of Advocate General Bobek delivered on 18 January 2018(1)," Case C-528/16, accessed June 6, 2018, http://curia. europa.eu/juris/document/document.jsf?text=&docid=198532&pageIndex=0&do clang=EN&mode=req&dir=&occ=first&part=1&cid=779174. This EU decision was then overturned by Europe's highest court, which ruled in July 2018 that crops altered by gene-editing techniques like CRISPR, even when no foreign DNA was added, would be regulated with the same stringency as GMOs. "Organisms Obtained by Mutagenesis Are GMOs and Are, in Principle, Subject to the Obligations Laid Down by the GMO Directive," Court of Justice of the European Union, press release no. 111/18, Luxembourg, July 25, 2018, Judgment in Case C-528/16, https://curia.europa.eu/jcms/upload/docs /application/pdf/2018–07/cp180111en.pdf.

9 Vitamin A has been genetically engineered into maize, cassava, and sweet potatoes; iron into beans and pearl millet; and zinc into rice and wheat. See Heather Ohly and Nicola Lowe, "Scientists Are Breeding Super-Nutritious Crops to Help Solve Global Hunger," *The Conversation,* June 1, 2018, accessed June 3, 2018, https://theconversation.com /scientists-are-breeding-super-nutritious-crops-to-help-solve-global-hunger-89815.

10 Bill Gates, "Gene Editing for Good," *Foreign Affairs* (May/June 2018), accessed May 3, 2018, https://www.foreignaffairs.com/articles/2018-04-10/gene-editing-good.

11 Tom Whipple, "Bill Gates Pumps Millions into Quest for a Supercow," *The Times,* January 26, 2018, https://www.thetimes.co.uk/article/bill-gates-pumps-millions-into-quest-for -a-supercow-7swc6dntw.

12 Sara Reardon, "Welcome to the CRISPR Zoo," *Nature* 531 (2016): 160–163, https:// www.nature.com/news/welcome-to-the-crispr-zoo-1.19537. BGI Shenzhen announced in 2017 it was dropping its micropig program, likely because the press attention it was getting detracted from the company's initial public offering.

13 For a fun description of this process, see Ben Mezrich, *Woolly: The True Story of the Quest to Revive One of History's Most Iconic Extinct Creatures* (New York: Simon & Schuster, 2018); and also, George Church and Ed Regis, *Regenesis* (New York: Basic Books, 2014): 10–11.

14 Barbara Sibbald, "Death But One Unintended Consequence of Gene-Therapy Trial," *Canadian Medical Association Journal* 164 (2001): 1612, https://www.ncbi.nlm.nih.gov /pmc/articles/PMC81135/; Osagie K. Obasogie, "Ten Years Later: Jesse Gelsinger's Death and Human Subjects Protection," The Hastings Center, October 22, 2009, https://www .thehastingscenter.org/ten-years-later-jesse-gelsingers-death-and-human-subjects-protection/.

15 Roland W. Herzog, Ou Cao, and Arun Srivastava, "Two Decades of Clinical Gene Therapy—Success Is Finally Mounting," *Discovery Medicine* 9 (2010): 105–11, https:// www.ncbi.nlm.nih.gov/pmc/articles/PMC3586794/.

16 Herzog, Cao, and Srivastava, "Two Decades of Clinical Gene Therapy."

17 Calvin J. Stephens et al., "Targeted In Vivo Knock-in of Human Alpha-1-antitrypsin CDNA Using Adenoviral Delivery of CRISPR/Cas9," *Gene Therapy* 25, no. 2 (2018): 139–56. doi:10.1038/s41434-018-0003-1.

18 Francis S. Collins and Scott Gottlieb, "The Next Phase of Human Gene-Therapy

Oversight," *New England Journal of Medicine* (August 15, 2018): doi:10.1056/nej mp1810628.

19 H-Y Xue et al., "*In Vivo* Gene Therapy Potentials of CRISPR-Cas9," *Gene Therapy* 23 (2016): 557–559, https://www.nature.com/articles/gt201625.

20 Lukas Villiger et al., "Treatment of a Metabolic Liver Disease by in Vivo Genome Base Editing in Adult Mice," *Nature Medicine* 24, no. 10 (October 2018): 1519–525, doi:10.1038/s41591-018-0209-1.

21 Collins and Gottlieb, "The Next Phase of Human Gene-Therapy Oversight."

22 Megan Molteni, "Biology Will Be the Next Great Computing Platform," *Wired*, May 3, 2018, accessed May 4, 2018, https://www.wired.com/story/biology-will-be-the-next -great-computing-platform/.

23 Gerald Schwank et al., "Functional Repair of CFTR by CRISPR/Cas9 in Intestinal Stem Cell Organoids of Cystic Fibrosis Patients," *Cell Stem Cell* 13 (2013): 653–658, http:// www.cell.com/cell-stem-cell/fulltext/S1934–5909(13)00493–1.

24 Hao Yin et al., "Genome Editing with Cas9 in Adult Mice Corrects a Disease Mutation and Phenotype," *Nature Biotechnology* 32 (2014): doi:10.1038/nbt.2884; Yanjiao Shao et al., "Cas9-Nickase–Mediated Genome Editing Corrects Hereditary Tyrosinemia in Rats," *Journal of Biological Chemistry* 293, no. 18 (2018): 6883–892. doi:10.1074/jbc .ra117.000347.

25 Puping Liang et al., "CRISPR/Cas9-Mediated Gene Editing in Human Tripronuclear Zygotes," *Protein & Cell* 6 (2015): 363–372, http://www.ncbi.nlm.nih.gov/pubmed /25894090.

26 Rafal Kaminski, "Elimination of HIV-1 Genomes from Human T-lymphoid Cells by CRISPR/Cas9 Gene Editing," *Scientific Reports* 6 (2016).

27 Hong Ma, "Correction of a Pathogenic Gene Mutation in Human Embryos," *Nature* 548 (2017): 413–419.

28 Chengzu Long et al., "Correction of Diverse Muscular Dystrophy Mutations in Human Engineered Heart Muscle by Single-Site Genome Editing," *Science Advances* 4 (2018), http://advances.sciencemag.org/content/4/1/eaap9004.

29 Alice Park, "Food and Drug Administration Approves a New Way to Use Gene Therapy," *Time*, December 20, 2017, accessed June 17, 2018. http://time.com/5073751/gene -therapy-visual-impairment/.

30 Serena H. Chen et al., "A Limited Survey-Based Uncontrolled Follow-Up Study of Children Born after Ooplasmic Transplantation in a Single Centre." *Reproductive BioMedicine Online* 33, no. 6 (2016): 737–44, doi:10.1016/j.rbmo.2016.10.003.

31 Rosa J. Castro, "Mitochondrial Replacement Therapy: The UK and U.S. Regulatory Landscapes," *Journal of Law and the Biosciences* 3 (2016): 726–35, https://academic.oup .com/jlb/article/3/3/726/2566730.

32 Although information about the UK mitochondrial transfer embryos is being withheld from the public until the babies are born, the 2019 date is both a logical inference based on when the treatments began to be administered in 2018 and on private correspondence with parties directly involved in this process. Because the final edits of Hacking Darwin were submitted in November 2018, it was not possible to include the exact birth date of the first UK mitochondrial transfer baby in this edition of the book. Ian Sample, "UK Doctors

Select First Women to Have 'Three-Person Babies,'" *The Guardian*, February 1, 2018, https://www.theguardian.com/science/2018/feb/01/permission-given-to-create -britains-first-three-person-babies. Singapore is also considering whether to authorize clinical trials of mitochondrial transfer.

33   Darwin Life, https://www.darwinlife.com/. In August 2017, however, Zhang received a letter from the FDA telling him that his public offer to place the DNA of older women into the donor eggs of younger women "cannot legally be performed in the United States. Nor is exportation permitted." See Ariana Eunjung Cha, "This Fertility Doctor Is Pushing the Boundaries of Human Reproduction, with Little Regulation." *Washington Post*, May 14, 2018, accessed June 1, 2018, https://www.washingtonpost.com/national /health-science/this-fertility-doctor-is-pushing-the-boundaries-of-human-reproduction -with-little-regulation/2018/05/11/ea9105dc-1831-11e8-8b08-027a6ccb38eb_story .html?utm_term=.220a3ce466e2.

34   Rob Stein, "Clinic Claims Success in Making Babies With 3 Parents' DNA." NPR, June 06, 2018, accessed June 7, 2018, https://www.npr.org/sections/health-shots/2018/06/06/61 5909572/inside-the-ukrainian-clinic-making-3-parent-babies-for-women-who-are-infertile.

35   C. Sallevelt et al., "Preimplantation Genetic Diagnosis for Single Gene Disorders," *Journal of Medical Genetics* 50 (2013): 125–132, http://jmg.bmj.com/content/50/2/125.long.

36   Shoukhrat Mitalipov et al., "Limitations of Preimplantation Genetic Diagnosis for Mitochondrial DNA Diseases," *Cell Reports* 7, no. 4 (2014): 935–37, doi:10.1016 /j.celrep.2014.05.004.

37   "Mitochondrial Replacement Therapy," United Mitochondrial Disease Foundation, https://www.umdf.org/mitochondrial-replacement-therapy/.

38   Rachel Kahn Best, "Disease Politics and Medical Research Funding," *American Sociological Review* 77 (2012): 780–803, http://journals.sagepub.com/doi/10.1177/0003122412 458509; Sara Reardon, "Lobbying Sways NIH Grants." *Nature* 515 (November 2014): 19.

39   David Cyranoski and Sara Reardon, "Chinese Scientists Genetically Modify Human Embryos," *Nature*, April 22, 2015, https://www.nature.com/news/chinese-scientists -genetically-modify-human-embryos-1.17378.

40   Ewen Callaway, "Second Chinese Team Reports Gene Editing in Human Embryos," *Nature*, April 8, 2016, https://www.nature.com/news/second-chinese-team-reports -gene-editing-in-human-embryos-1.19718#/b2.

41   Hong Ma et al., "Correction of a Pathogenic Gene Mutation in Human Embryos," *Nature* 548 (2017): 413–419, https://www.nature.com/articles/nature23305.

42   Pam Belluck, "In Breakthrough, Scientists Edit a Dangerous Mutation from Genes in Human Embryos," *New York Times*, August 2, 2017, https://www.nytimes .com/2017/08/02/science/gene-editing-human-embryos.html.

CHAPTER 6

1   Flannick et al., "Loss-of-Function Mutations in SLC30A8 Protect against Type 2 Diabetes," *Nature Genetics* 46 (2014): 357–363, https://www.ncbi.nlm.nih.gov /pubmed/24584071.

2   Rong Liu et al., "Homozygous Defect in HIV-1 Coreceptor Accounts for Resistance

of Some Multiply-Exposed Individuals to HIV-1 Infection," *Cell* 86 (1996): 367–377, http://www.cell.com/abstract/S0092–8674(00)80110–5.

3   Christopher P. Cannon et al., "Ezetimibe Added to Statin Therapy after Acute Coronary Syndromes," *New England Journal of Medicine* 372 (2015): 2387–2397, http://www.nejm .org/doi/full/10.1056/NEJMoa1410489.

4   Yanfang Fu et al., "High-Frequency Off-Target Mutagenesis Induced by CRISPR-Cas Nucleases in Human Cells," *Nature Biotechnology* 31 (2013): 822–826.

5   David Cyranoski and Sara Reardon, "Chinese Scientists Genetically Modify Human Embryos," *Nature*, April 22, 2015, https://www.nature.com/news/chinese-scientists -genetically-modify-human-embryos-1.17378.

6   Emma Haapaniemi et al., "CRISPR–Cas9 Genome Editing Induces a P53-mediated DNA Damage Response." *Nature Medicine*, 2018, doi:10.1038/s41591-018-0049-z.

7   Kosicki, Michael, Kärt Tomberg, and Allan Bradley. "Repair of Double-Strand Breaks Induced by CRISPR–Cas9 Leads to Large Deletions and Complex Rearrangements." *Nature Biotechnology* 36 (July 16, 2018): 765–771. The share price of CRISPR-related companies regularly dip when major reports like this are released. CRISPR Therapeutics, Editas Medicine, and Intellia Therapeutics lost a collective $300 million in share value the day this report was released in July 2018.

8   Emily Mullin, "CRISPR 2.0 Is Here, and It's Way More Precise," *MIT Technology Review,* October 25, 2017, https://www.technologyreview.com/s/609203/crispr-20-is-here -and-its-way-more-precise/.

9   Mullin, "CRISPR 2.0 Is Here, and It's Way More Precise."

10   Nicole M. Gaudelli et al., "Programmable Base Editing of A•T to G•C in Genomic DNA without DNA Cleavage," *Nature* 551 (2017): 464–471, http://evolve.harvard.edu/138 -ABE.pdf.

11   Yanting Zeng et al., "Correction of the Marfan Syndrome Pathogenic FBN1 Mutation by Base Editing in Human Cells and Heterozygous Embryos," *Molecular Therapy*, August 13, 2018. doi:10.1016/j.ymthe.2018.08.007.

12   Michael Gapinske et al., "CRISPR-SKIP: Programmable Gene Splicing with Single Base Editors," *Genome Biology* 19, no. 1 (2018). doi:10.1186/s13059-018-1482-5.

13   Cassandra Willyard, "The Epigenome Editors: How Tools Such as CRISPR Offer New Details about Epigenetics," *Nature Medicine* 23, no. 8 (2017): 900–03, doi:10.1038 /nm0817-900; Ianis G. Matsoukas, "Commentary: RNA Editing with CRISPR-Cas13," *Frontiers in Genetics* 9 (2018), doi:10.3389/fgene.2018.00134.

14   Zhuchi Tu et al., "Promoting Cas9 Degradation Reduces Mosaic Mutations in Non-Human Primate Embryos," *Scientific Reports* 7 (2017), https://www.nature .com/articles/srep42081; Michael Le Page, "Mosaic Problem Stands in the Way of Gene Editing Embryos," *New Scientist*, March 15, 2017, https://www.newscientist .com/article/mg23331174–400-mosaic-problem-stands-in-the-way-of-gene-editing -embryos/; P. Singh, J. C. Schimenti, and E. Bolcun-Filas, "A Mouse Geneticist's Practical Guide to CRISPR Applications," *Genetics* 199 (2014): 1–15, http://www.genetics .org/content/199/1/1.full; Michael Le Page, "Male Infertility Cure Will Be Gateway to Editing Our Kids' Genes," *New Scientist,* June 23, 2016, https://www.newscientist.com /article/2094926-male-infertility-cure-will-be-gateway-to-editing-our-kids-genes/.

15   Hong Ma et al., "Correction of a Pathogenic Gene Mutation in Human Embryos," *Nature* 548 (2017): 413–419, https://www.nature.com/articles/nature23305.

16   Dieter Egli et al., "Inter-homologue Repair in Fertilized Human Eggs?" *BioRxiv* (2017).

17   Adikusuma, Fatwa, Sandra Piltz, Mark A. Corbett, Michelle Turvey, Shaun R. Mccoll, Karla J. Helbig, Michael R. Beard, James Hughes, Richard T. Pomerantz, and Paul Q. Thomas. "Large Deletions Induced by Cas9 Cleavage," *Nature* 560, no. 7,717 (2018). doi:10.1038/s41586-018-0380-z; Dieter Egli, et al. "Inter-homologue Repair in Fertilized Human Eggs?" *Nature* 560 (2018); Fatwa Adikusuma et al., "Large Deletions Induced by Cas9 Cleavage," *Nature* 560 (August 2018): E8–E9.

18   Genome Editing and Human Reproduction. Report. July 17, 2018, Nuffield Council on Bioethics. http://nuffieldbioethics.org/wp-content/uploads/Genome-editing-and -human-reproduction-FINAL-website.pdf, p. 45.

19   Dennis Normile, "CRISPR Bombshell: Chinese Researcher Claims to Have Created Gene-Edited Twins," *Science*, November 26, 2018, https://www.sciencemag.org /news/2018/11/crispr-bombshell-chinese-researcher-claims-have-created-gene-edited -twins.

20   Jacob Gratten and Peter M. Visscher, "Genetic Pleiotropy in Complex Traits and Diseases: Implications for Genomic Medicine," *Genome Medicine* 8 (2016), https://www.ncbi.nlm .nih.gov/pmc/articles/PMC4952057/.

21   Evan A. Boyle et al., "An Expanded View of Complex Traits: From Polygenic to Omnigenic," *Cell* 169 (2017): 1177–1186, http://www.cell.com/cell/fulltext/S0092 –8674(17)30629-3.

22   The omnigenic hypothesis remains controversial and has been hotly debated by scientists since the publication of the 2017 Boyle and Li article in *Cell*. See "The Omnigenic Model: Special Issue," *Journal of Psychiatry and Brain Science* 2, no. 5 (2017). doi:10.20900/jpbs.20170014s1, http://jpbs.qingres.com/IssueList.aspx?years no=2017&volumeno=2&issueno=5 and Naomi Wray et al., "Common Disease Is More Complex Than Implied by the Core Gene Omnigenic Model," *Cell* 173, no. 7 (June 14, 2018): 1573–1580, doi:10.1016/j.cell.2018.05.051.

23   I hope you will visit the website of this incredible project: http://www.openworm.org/.

24   Lucy Black, "A Worm's Mind in a Lego Body," *I Programmer*, November 16, 2014, http:// www.i-programmer.info/news/105-artificial-intelligence/7985-a-worms-mind-in-a -lego-body.html#.

25   Human Cell Atlas, https://www.humancellatlas.org/.

26   Ray Kurzweil, "The Law of Accelerating Returns," Kurzweil Accelerating Intelligence, March 7, 2001, http://www.kurzweilai.net/the-law-of-accelerating-returns.

27   Ray Kurzweil, *The Singularity Is Near: When Humans Transcend Biology* (London: Duckworth, 2016): 39. For a wonderful explanation, see also Tim Urban, "The Artificial Intelligence Revolution: The Road to Superintelligence," Wait But Why, January 22, 2015, https://waitbutwhy.com/2015/01/artificial-intelligence-revolution-1.html.

28   Homer, *The Iliad*, trans. Robert Fagles (Chicago: Penguin Books, 1951).

29   Dante Aligheri, *Divine Comedy*, trans. Henry Wadsworth Longfellow (Project Gutenburg EBook, 1997): Canto 17.

30 Catherine Easterbrook and Guy Maddern, "Porcine and Bovine Surgical Products: Jewish, Muslim, and Hindu Perspectives," *JAMA* 143 (2008): 366–370, https://jamanetwork .com/journals/jamasurgery/fullarticle/599037.

31 Simon Bramhall, "Presumed Consent for Organ Donation: A Case Against," *Annals of the Royal College of Surgeons of England* 93 (2011): 270–272, https://www.ncbi.nlm. nih.gov/pmc/articles/PMC3363073/. Monty Python does a hilarious spoof of "live organ transplants" in its 1983 film, *The Meaning of Life*. https://www.youtube.com/ watch?v=5ig9wr8517E.

32 Policy changes, like shifting from an opt-out to an opt-in system for organ donation or allowing compensation to be provided to the families of donors, could help address this shortfall. "Organ Donation Statistics," U.S. Department of Health and Human Services, https://www.organdonor.gov/statistics-stories/statistics.html.

33 Bethany Pellegrino, "Immunosuppression," *Medscape*, last modified January 4, 2016, https://emedicine.medscape.com/article/432316-overview#a2.

34 Li Wei et al., "Inactivation of Porcine Endogenous Retrovirus in Pigs Using CRISPR-Cas9," *Science*, September 22, 2017. http://science.sciencemag.org/content/357/6357/1303.

35 Nicola Davis, "Breakthrough as Scientists Grow Sheep Embryos Containing Human Cells," *The Guardian*, February 17, 2018, https://www.theguardian.com/science/2018 /feb/17/breakthrough-as-scientists-grow-sheep-embryos-containing-human-cells.

36 Registry of Standard Biological Parts, http://parts.igem.org/Main_Page.

37 The Free Genes Project, https://biobricks.org/freegenes/.

38 Onkar Sumant, "Synthetic Biology Market by Products (DNA Synthesis, Oligonucleotide Synthesis, Synthetic DNA, Synthetic Genes, Synthetic Cells, XNA) and Technology (Genome Engineering, Microfluidics Technologies, DNA Synthesis & Sequencing Technologies): Global Opportunity Analysis and Industry Forecast, 2013–2020," March 2014, https://www.alliedmarketresearch.com/synthetic-biology-market; "The Global Synthetic Biology Market Is Projected to Grow at a CAGR of 19.9%," *PR Newswire*, January 5, 2018, https://www.prnewswire.com/news-releases/the-global-synthetic -biology-market-is-projected-to-grow-at-a-cagr-of-199–300578132.html.

39 Steven Cerier, "Synthetic Biology's 'Promise and Potential' Capture Investor Attention." Genetic Literacy Project, May 8, 2018, accessed May 8, 2018, https:// geneticliteracyproject.org/2018/05/08/synthetic-biologys-promise-potential-investor -attention/?mc_cid=9b36f13d19&mc_eid=6d7f502b6d.

40 Robert F. Service, "Synthetic Microbe Lives with Fewer Than 500 Genes," *Science*, December 9, 2017, accessed May 8, 2018, http://www.sciencemag.org/news/2016/03 /synthetic-microbe-lives-fewer-500-genes.

41 David Ewing Duncan, "Is the World Ready for Synthetic People?" Neo.Life, April 5, 2018, accessed April 22, 2018, https://medium.com/neodotlife/q-a-with-drew-endy -bde0950fd038. See also George M. Church and Edward Regis, *Regenesis: How Synthetic Biology Will Reinvent Nature and Ourselves* (New York: Basic Books, 2014): 53.

42 Importantly, the HGP-write scientists stressed the ethical consideration of this initiative. Jef D. Boeke et al., "The Genome Project-Write," *Science* 353 no. 6295 (July 8, 2016): 126–127, http://science.sciencemag.org/content/353/6295/126.

43   David Ewing Duncan, "The Next Best Version of Me: How to Live Forever," *Wired*, March 27, 2018, https://www.wired.com/story/live-forever-synthetic-human-genome/.

44   The Mason Lab, http://www.masonlab.net/.

CHAPTER 7

1   David Ferry, *Gilgamesh: A New Rendering in English Verse* (New York: Farrar, Straus, and Giroux, 1992): 56–57.

2   Eric Grundhauser, "The True Story of Dr. Voronoff's Plan to Use Monkey Testicles to Make Us Immortal," *Atlas Obscura,* October 13, 2015, https://www.atlasobscura.com /articles/the-true-story-of-dr-voronoffs-plan-to-use-monkey-testicles-to-make-us-immortal.

3   Renee Stepler, "World's Centenarian Population Projected to Grow Eightfold by 2050," Pew Research Center, April 21, 2016, http://www.pewresearch.org/fact-tank/2016/04/21 /worlds-centenarian-population-projected-to-grow-eightfold-by-2050/.

4   "Living to 120 and Beyond: Americans' Views on Aging, Medical Advances and Radical Life Extension," Pew Research Center, August 6, 2013, http://www.pewforum .org/2013/08/06/living-to-120-and-beyond-americans-views-on-aging-medical -advances-and-radical-life-extension/.

5   Xian Xia et al., "Molecular and Phenotypic Biomarkers of Aging," *F1000 Research* 6 (2017), https://www.ncbi.nlm.nih.gov/pmc/articles/PMC5473407/.

6   Carola Weidner, "Aging of Blood Can Be Tracked by DNA Methylation Changes at Just Three CpG Sites," *Genome Biology* 15 (2014), https://www.ncbi.nlm.nih.gov/pmc /articles/PMC4053864/pdf/gb-2014-15-2-r24.pdf. Interestingly, the p53 gene has also been shown to play an important cancer-fighting role in elephants. See Carl Zimmer, "The 'Zombie Gene' That May Protect Elephants from Cancer," *New York Times*, August 14, 2018, https://www.nytimes.com/2018/08/14/science/the-zombie-gene-that-may -protect-elephants-from-cancer.html.

7   Masayuki Kimura et al., "Telomere Length and Mortality: A Study of Leukocytes in Elderly Danish Twins," *American Journal of Epidemiology* 167 (2008): 799–806, https:// www.ncbi.nlm.nih.gov/pmc/articles/PMC3631778/.

8   F. Huber et al., "Walking Speed as an Aging Biomarker in Baboons (Papio Hamadryas)." *Journal of Medical Primatology*, U.S. National Library of Medicine (December 2015), https://www.ncbi.nlm.nih.gov/pmc/articles/PMC4802968/.

9   Weiyang Chen et al., "Three-Dimensional Human Facial Morphologies as Robust Aging Markers," *Cell Research* 25, no. 5 (May 2015): 574–587, accessed April 28, 2018, https:// www.ncbi.nlm.nih.gov/pmc/articles/PMC4423077/.

10   A group of revisionist anthropologists recently made the case that grandmothers in particular play a more important evolutionary role than previously imagined. Matthew H. Chan, Kristen Hawkes, and Peter S. Kim. "Modelling the Evolution of Traits in a Two-Sex Population, with an Application to Grandmothering," *Bulletin of Mathematical Biology* 79, no. 9 (2017): 2132–148, doi:10.1007/s11538-017-0323-0.

11   Julie A. Mattison et al., "Caloric Restriction Improves Health and Survival of Rhesus Monkeys," *Nature Communications* 8 (2017), https://www.nature.com/articles

/ncomms14063. Although at first it looked like one of the studies showed that CR worked in extending the life and health of the monkeys and the other did not, a later review of the data confirmed that both studies shared the same result.

12 Leanne M. Redman et al., "Metabolic Slowing and Reduced Oxidative Damage with Sustained Caloric Restriction Support the Rate of Living and Oxidative Damage Theories of Aging," *Cell Metabolism* 27 (2018): 1–11.

13 These chronic diseases account for over 90 percent of all deaths in the United States. Khadija Ismail et al., "Compression of Morbidity Is Observed Across Cohorts with Exceptional Longevity," *Journal of the American Geriatrics Society* 64 (2016): 1583–1591, https://www.ncbi.nlm.nih.gov/pubmed/27377170.

14 "Wellderly Study Suggests Link Between Genes That Protect against Cognitive Decline and Overall Healthy Aging," The Scripps Research Institute, April 21, 2016, https://www.scripps.edu/news/press/2016/20160421wellderly.html; Galina A. Erikson et al., "Whole-Genome Sequencing of a Healthy Aging Cohort," *Cell* 165 (2016): 1002–1011, http://www.cell.com/cell/fulltext/S0092-8674(16)30278-1.

15 Freudenberg-Hua et al., "Disease Variants in Genomes of 44 Centenarians," *Molecular Genetics and Genomic Medicine* 2 (2014): 438–450, https://www.ncbi.nlm.nih.gov/pubmed/25333069.

16 Clyde B. Schechter et al., "Cholesteryl Ester Transfer Protein (CETP) Genotype and Reduced CETP Levels Associated with Decreased Prevalence of Hypertension," *Mayo Clinic Proceedings* 85 (2010): 522–526, https://www.ncbi.nlm.nih.gov/pubmed/20511482.

17 Nir Barzilai and Ilan Gabriely, "Genetic Studies Reveal the Role of the Endocrine and Metabolic Systems in Aging," *The Journal of Clinical Endocrinology & Metabolism* 95 (2010): 4493–4500, https://www.ncbi.nlm.nih.gov/pmc/articles/PMC3050096/.

18 Galina A. Erikson et al., "Whole-Genome Sequencing of a Healthy Aging Cohort," *Cell* 165 (2016): 1002–1011, http://www.cell.com/cell/fulltext/S0092-8674(16)30278-1.

19 Luke C. Pilling et al., "Human Longevity: 25 Genetic Loci Associated in 389,166 UK Biobank Participants," *Aging* 9 (2017): 2504–2520, http://www.aging-us.com/article/101334/text#fulltext.

20 Dan Buettner, *The Blue Zones: Lessons for Living Longer from the People Who've Lived the Longest* (National Geographic Books, 2009): vii.

21 R. Speakman, "Body Size, Energy Metabolism and Life Span," *Journal of Experimental Biology* 208 (2005): 1717–1730, https://www.ncbi.nlm.nih.gov/pubmed/15855403.

22 Antonio Regalado, "Google's Long, Strange, Life Span Trip," *MIT Technology Review*, December 15, 2016, https://www.technologyreview.com/s/603087/googles-long-strange-life-span-trip/.

23 Karl A. Rodriguez et al., "Determinants of Rodent Longevity in the Chaperone-protein Degradation Network," *Cell Stress and Chaperones* 21, no. 3 (2016): 453–66. doi:10.1007/s12192-016-0672-x.

24 Jorge Azpurua et al., "Naked Mole-Rat Has Increased Translational Fidelity Compared with the Mouse, as well as a Unique 28S Ribosomal RNA Cleavage," *Proceedings of the National Academy of Sciences of the United States of America* 110 (2013): 17350–17355, http://www.pnas.org/content/110/43/17350.

25   Joseph Stromberg, "Why Do Naked Mole Rats Live So Long?" *Smithsonian.com*, September 30, 2013, https://www.smithsonianmag.com/science-nature/why-do-naked -mole-rats-live-so-long-230258/.

26   Ungvari et al., "Extreme Longevity Is Associated with Increased Resistance to Oxidative Stress in *Arctica islandica*, the Longest-Living Non-Colonial Animal," *The Journals of Gerontology Series A: Biological Sciences and Medical Sciences* 66A(2011): 741–750, https://www.ncbi.nlm.nih.gov/pmc/articles/PMC3143345/.

27   Piraino et al., "Reversing the Life Cycle: Medusae Transforming into Polyps and Cell Transdifferentiation in Turritopsis nutricula (Cnidaria, Hydrozoa)," *The Biological Bulletin* 190 (1996): 302–312, https://www.jstor.org/stable/1543022?seq=1#page_scan_tab _contents.

28   For a fantastic review of how scientists are studying various long-lived simple organisms, see: Ronald S. Petralia, Mark P. Mattson, and Pamela J. Yao, "Aging and Longevity in the Simplest Animals and the Quest for Immortality," *Ageing Research Reviews* 16 (2014): 66–82, https://www.ncbi.nlm.nih.gov/pmc/articles/PMC4133289/.

29   The mutation of a single gene called *Daf-2* helped the roundworms live twice as long as their compatriots without the mutation. A mutation on the Daf-16 gene had the opposite effect. It is important to note, however, that the life span extending mutations also had significant side effects including decreased fertility, smaller size, and sometimes a higher chance of dying in utero. Cynthia J. Kenyon, "The Genetics of Ageing," *Nature* 464 (2010): 504–512, https://www.nature.com/articles/nature08980; Masaharu Uno and Eisuke Nishida, "Life Span-Regulating Genes in *C. elegans*," *Aging and Mechanisms of Disease* 2 (2016); "Caenorhabditis Elegans," AnAge: The Animal Ageing and Longevity Database, http://genomics.senescence.info/species/entry.php?species=Caenorhabditis_Elegans; David Michaelson et al., "Insulin Signaling Promotes Germline Proliferation in C. Elegans." Development, February 15, 2010, accessed May 12, 2018, https://www.ncbi .nlm.nih.gov/pmc/articles/PMC2827619/.

30   Elizabeth H. Blackburn and Elissa Epel, *The Telomere Effect: A Revolutionary Approach to Living Younger, Healthier, Longer* (London: Orion Spring, 2018). See also my conversation with Nobel laureate and telomere pioneer Elizabeth Blackburn (and Harvard geneticist George Church) at the 2017 Google Zeitgeist conference, Zeitgeistminds, "Unlocking the Code of Life," YouTube, October 24, 2017, accessed May 25, 2018, https://www.youtube .com/watch?v=srX79RA-HPQ&feature=youtu.be.

31   Wei et al., "Fasting-Mimicking Diet and Markers/Risk Factors for Aging, Diabetes, Cancer, and Cardiovascular Disease," *Science Translational Medicine* 9 (2017), http://stm .sciencemag.org/content/9/377/eaai8700.short.

32   Steven C. Moore et al., "Leisure Time Physical Activity of Moderate to Vigorous Intensity and Mortality: A Large Pooled Cohort Analysis," *PLoS Medicine* 9 (2012), http:// journals.plos.org/plosmedicine/article?id=10.1371/journal.pmed.1001335.

33   Michael S. Bonkowski and David A. Sinclair, "Slowing Ageing by Design: The Rise of NAD and Sirtuin-Activating Compounds," *Nature Reviews Molecular Cell Biology* 17 (2016): 679–690, https://www.nature.com/articles/nrm.2016.93; Abhirup Das et al., "Impairment of an Endothelial NAD-H 2 S Signaling Network Is a Reversible Cause of Vascular Aging," *Cell* 173 (2018): 74–89.

34   M. Evans et al., "Metformin and Reduced Risk of Cancer in Diabetic Patients," *BMJ* 330 (2005): 1304–1305, https://www.ncbi.nlm.nih.gov/pubmed/15849206.

35   A. Bannister et al., "Can People with Type 2 Diabetes Live Longer Than Those Without? A Comparison of Mortality in People Initiated with Metformin or Sulphonylurea Monotherapy and Matched, Non-Diabetic Controls," *Diabetes, Obesity and Metabolism* 16 (2014): 1165–1173, http://onlinelibrary.wiley.com/doi/10.1111/dom.12354/full.

36   P. Ng et al., "Long-Term Metformin Usage and Cognitive Function among Older Adults with Diabetes," *Journal of Alzheimer's Disease* 41 (2014): 61–68, https://www.ncbi.nlm.nih.gov/pubmed/24577463; V. N. Anisimov, "Metformin for Cancer and Aging Prevention: Is It a Time to Make the Long Story Short?" *Oncotarget*. November 24, 2015, accessed April 28, 2018, https://www.ncbi.nlm.nih.gov/pubmed/26583576. Other studies have shown a lesser positive impact of metformin.

37   Alejandro Martin-Montalvo, "Metformin Improves Healthspan and Life Span in Mice," *Nature Communications* 4 (2013), https://www.nature.com/articles/ncomms3192.

38   Jing Li, Sang Gyun Kim, and John Blenis, "Rapamycin: One Drug, Many Effects," *Cell Metabolism* 19 (2014): 373–379, https://www.ncbi.nlm.nih.gov/pmc/articles/PMC3972801/.

39   Simon C. Johnson, Peter S. Rabinovitch, and Matt Kaeberlein, "MTOR Is a Key Modulator of Ageing and Age-Related Disease," *Nature* 493 (2013): 338–345, https://www.ncbi.nlm.nih.gov/pmc/articles/PMC3687363/.

40   Dog Aging Project, accessed June 18, 2018, http://dogagingproject.com/. Silvan R. Urfer et al., "A Randomized Controlled Trial to Establish Effects of Short-Term Rapamycin Treatment in 24 Middle-Aged Companion Dogs," *GeroScience*39, no. 2 (2017): 117–27. doi:10.1007/s11357-017-9972-z; Matt Kaeberlein, Kate E. Creevy, and Daniel E. L. Promislow, "The Dog Aging Project: Translational Geroscience in Companion Animals," *Mammalian Genome* 27, no. 7–8 (2016): 279–88, doi:10.1007/s00335-016-9638-7.

41   Simon C. Johnson, Peter S. Rabinovitch, and Matt Kaeberlein, "MTOR Is a Key Modulator of Ageing and Age-Related Disease," *Nature* 493 (2013): 338–345, https://www.ncbi.nlm.nih.gov/pmc/articles/PMC3687363/.

42   Simon C. Johnson and Matt Kaeberlein, "Rapamycin in Aging and Disease: Maximizing Efficacy While Minimizing Side Effects," *Oncotarget*7 (2016): 44876–44878, https://www.ncbi.nlm.nih.gov/pmc/articles/PMC5216691/; V. N. Anisimov, "Metformin for Cancer and Aging Prevention: Is It a Time to Make the Long Story Short?" *Oncotarget* November 24, 2015, accessed April 28, 2018, https://www.ncbi.nlm.nih.gov/pubmed/26583576.

43   A series of recent mouse studies point in the direction of how this type of cocktail might work. In one, metformin increased the life span of mice by 7 percent overall, but mice fed both metformin and rapamycin together were the longest-lived mice studied. Female mice fed a daily dose of rapamycin alone lived 18 to 21 percent longer than nondrugged mice, but female mice fed rapamycin and metformin together lived 23 percent longer. Males lived 10 to 13 percent longer with rapamycin alone but 23 percent longer with both. L. J. Wei, D. Y. Lin, and L. Weissfeld, "Regression Analysis of Multivariate Incomplete Failure Time Data by Modeling Marginal Distributions," *Journal of the American Statistical Association* 84 (1989): 1065, http://dlin.web.unc.edu/files/2013/04/WeiLinWeissfeld89.pdf.

44   Abel Soto-Gamez and Marco Demaria, "Therapeutic Interventions for Aging: The

Case of Cellular Senescence," *Drug Discovery Today* 22 (2017): 786–795, http://www
.sciencedirect.com/science/article/pii/S135964461730017X.

45   Eva Latorre et al., "Mitochondria-Targeted Hydrogen Sulfide Attenuates Endothelial
Senescence by Selective Induction of Splicing Factors HNRNPD and SRSF2," *Aging*,
August 19, 2018. doi:10.18632/aging.101500.

46   P. Baar et al., "Target Apoptosis of Senescent Cells Restores Tissue Homeostasis in
Response to Chemotoxicity and Aging," *Cell* 169 (2017): 132–147, https://www.ncbi
.nlm.nih.gov/pubmed/28340339.

47   Alejandro Ocampo, "In Vivo Amelioration of Age-Associated Hallmarks by Partial
Reprogramming," *Cell* 167 (2016): 1719–1733, http://www.cell.com/cell/fulltext
/S0092–8674(16)31664–6.

48   "Turning Back Time: Salk Scientists Reverse Signs of Aging," The Salk Institute, December
15, 2016, https://www.salk.edu/news-release/turning-back-time-salk-scientists-reverse
-signs-aging/.

49   M. Conboy, "Rejuvenation of Aged Progenitor Cells by Exposure to a Young Systemic
Environment," *Nature* 433 (2005): 760–764, https://www.ncbi.nlm.nih.gov/pubmed
/15716955.

50   Jocelyn Kaiser, "Young Blood Renews Old Mice," *Science*, May 4, 2014, http://www
.sciencemag.org/news/2014/05/young-blood-renews-old-mice.

51   "Longevity Industry Landscape Overview 2017," Geroscience, Policy, and Economics,
p. 7, https://daks2k3a4ib2z.cloudfront.net/581ba14cc9b0d76c5dcededf/5a0e1db85
db6530001540224_Infographic%20Summary%20Longevity%20Industry%20Report.pdf.

52   P. Goldman et al., "Substantial Health And Economic Returns from Delayed Aging
May Warrant A New Focus for Medical Research," *Health Affairs* 32 (2013): 1698–
1705, http://sjayolshansky.com/sjo/Longevity_Dividend_Initative_files/Health%20
Affairs%202013%20LDI%20Final.pdf.

CHAPTER 8

1   Andrew Hammond et al., "A CRISPR-Cas9 Gene Drive System Targeting Female
Reproduction in the Malaria Mosquito Vector *Anopheles gambiae*," *Nature Biotechnology*
34 (2016): 78–83, https://www.nature.com/articles/nbt.3439. For a good description of
gene drives, see "FAQs: Gene Drives," Wyss Institute, Harvard University, accessed May
12, 2018, https://wyss.harvard.edu/staticfiles/newsroom/pressreleases/Gene drives
FAQ FINAL.pdf.

2   Peter, Paul and Mary, vocalists, "I Know an Old Lady," by Rose Bonne and Alan Mills,
1952, released 1993, track 7 on *Peter, Paul & Mommy, Too*, http://www.peterpaulandmary.
com/music/17–07.htm.

3   *Human Dignity and Bioethics: Essays Commissioned by the President's Council on Bioethics*
(Washington, DC: March 2008): 329, https://bioethicsarchive.georgetown.edu/pcbe
/reports/human_dignity/.

4   Michael J. Sandel, "The Case Against Perfection," *The Atlantic*, April 2014, https://www
.theatlantic.com/magazine/archive/2004/04/the-case-against-perfection/302927/.

5   Excerpt from letter 49. TO J.D. HOOKER. July 13th, 1856, More Letters of Charles

Darwin, Volume I, https://charles-darwin.classic-literature.co.uk/more-letters-of
-charles-darwin-volume-i/ebook-page-57.asp.

6   Marcy Darnovsky, "A Slippery Slope to Human Germline Modification," *Nature* 499, no.
     7457 (July 9, 2013): 127, http://www.nature.com/news/a-slippery-slope-to-human
     -germline-modification-1.13358.

7   César Palacios-González, John Harris, and Giuseppe Testa, "Multiplex Parenting: IVG
     and the Generations to Come," *Journal of Medical Ethics* 40 (March 7, 2014): 752–758,
     http://jme.bmj.com/content/early/2014/03/07/medethics-2013-101810.

8   Siddhartha Mukherjee, *The Gene: An Intimate History*. New York: Simon and Schuster,
     2016, pp. 275–276.

9   Julian Savulescu and Guy Kahane, "The Moral Obligation to Create Children with the
     Best Chance of the Best Life," *Bioethics* 23, no. 5 (2009): 274–290, https://www.ncbi
     .nlm.nih.gov/pubmed/19076124.

10   Nick Bostrom and Toby Ord, "How to Avoid Status Quo Bias in Bioethics: The Case for
     Human Enhancement," 2004, https://nickbostrom.com/ethics/statusquo.doc.

11   Carl Zimmer, *She Has Her Mother's Laugh: The Powers, Perversions, and Potential of Heredity*
     (New York: Dutton, 2018).

12   Rebecca Bennett, "When Intuition Is Not Enough. Why the Principle of Procreative
     Beneficence Must Work Much Harder to Justify Its Eugenic Vision," *Bioethics* 28 (July
     10, 2013): 447–455, http://onlinelibrary.wiley.com/doi/10.1111/bioe.12044/full;
     Biplab Kumar Halder, "Can the Principle of Procreative Beneficence Justify the Non-
     Medical Use of Preimplantation Genetic Diagnosis?" *Center for the Study of Ethics in
     Society Papers*, Paper 103 (2016), http://scholarworks.wmich.edu/cgi/viewcontent
     .cgi?article=1101&context=ethics_papers.

13   "The Laws of the Twelve Tables," c. 450 B.C.E., https://facultystaff.richmond
     .edu/~wstevens/FYStexts/twelvetables.pdf.

14   Francis Galton, *Inquiries into the Human Faculty and Its Development* (Macmillan,
     1883): 24.

15   *The Eugenics Review* 1, no. 1 (April 1909). https://www.ncbi.nlm.nih.gov/pmc/articles
     /PMC2990364/?page=1. The journal ceased publication only in 1968.

16   George F. Will, "The Liberals Who Loved Eugenics," *Washington Post*, March 8, 2017,
     https://www.washingtonpost.com/opinions/the-liberals-who-loved-eugenics/2017/03
     /08/0cc5e9a0-0362-11e7-b9fa-ed727b644a0b_story.html?utm_term=.5758cd0c98e3.

17   Henry Fairfield Osborn, *Collected Papers* (1877): Volume 4, p. 2. https://books.google
     .com/books?id=hHMuAAAAIAAJ&pg=PT32&lpg=PT32&dq=%22Negro+fails+in
     +government,+he+may+become+a+fine+agriculturist%22&source=bl&ots=OK7jUt
     TCeT&sig=cavOSWARMqiT550HwpTpopV9XXU&hl=en&sa=X&ved=2a
     hUKEwiMmc2PgOHcAhWO2FMKHWvFAVEQ6AEwAHoECAcQAQ#v=onep
     age&q=%22Negro%20fails%20in%20government%2C%20he%20may%20become%20
     a%20fine%20agriculturist%22&f=false.

18   Steven A. Farber, "U.S. Scientists' Role in the Eugenics Movement (1907–1939): A
     Contemporary Biologist's Perspective," *Zebrafish* 5, no. 4 (December 2008): 243–245,
     https://www.ncbi.nlm.nih.gov/pmc/articles/PMC2757926/.

19   For a list of materials on the particularly egregious case of Puerto Rico, see "Sterilization

of Puerto Rican Women: A Selected, Partially Annotated Bibliography (Louis de Malave, 1999)" Special Collections, March 23, 2018, accessed June 26, 2018, https://www.library .wisc.edu/gwslibrarian/publications/bibliographies/sterilization/.

20   Paul A. Lombardo, "Three Generations, No Imbeciles: New Light on *Buck v. Bell*," *NYU Law Review* 80 (1985): 30–62, https://pdfs.semanticscholar.org/784b /f1b7cfbbc84b6966f4c3b0f3d554726d551e.pdf.

21   Adolph Hitler, *Mein Kampf* (1925): 222–223, trans. Ralph Manheim (Boston: Houghton Mifflin Company, 1999). http://www.sjsu.edu/people/mary.pickering/courses /His146/s1/MeinKampfpartone0001.pdf.

22   Charles J. Epstein, "Is Modern Genetics the New Eugenics?" *Genetics in Medicine* 5 (2003): 469–475, https://www.nature.com/articles/gim2003376.

23   Michael J. Sandel, "The Case against Perfection," *The Atlantic*, April 2004, https://www .theatlantic.com/magazine/archive/2004/04/the-case-against-perfection/302927/.

24   Arthur Kaplan, "What Should the Rules Be?" *Time*, January 14, 2001, http://content .time.com/time/magazine/article/0,9171,95244,00.html.

25   Richard Dawkins, *The Genetic Revolution and Human Rights*, ed. Justine Burley (Oxford: Oxford University Press, 1999): v–xviii; cited in Charles J. Epstein, "Is Modern Genetics the New Eugenics?" *Nature*, November 1, 2003, accessed May 10, 2018, https://www .nature.com/articles/gim2003376.

26   Diane B. Paul, *The Politics of Heredity: Essays on Eugenics, Biomedicine, and the Nature-Nurture Debate* (New York: State University of NY Press, 1998) p. 97; cited in Charles J. Epstein, "Is Modern Genetics the New Eugenics?" *Nature*, November 1, 2003, accessed May 10, 2018, https://www.nature.com/articles/gim2003376.

27   Jon Entine, "Let's (Cautiously) Celebrate the 'New Eugenics,'" *Huffington Post*, October 30, 2014, https://www.huffingtonpost.com/jon-entine/lets-cautiously-celebrate_b_6070462.html.

28   Nicholas Agar, *Liberal Eugenics: In Defense of Human Enhancement* (Oxford: Blackwell, 2004): vi, 5. Nicholas Agar, "Liberal Eugenics," *Public Affairs Quarterly* 12 (1998): 137–155.

29   Adam Cohen, "Is There Such a Thing as Good Eugenics?" *Los Angeles Times*, March 17, 2017, http://www.latimes.com/opinion/op-ed/la-oe-cohen-good-eugenics-20170 317-story.html.

30   Stephen Whyte, Benno Torgler, and Keith L. Harrison, "What Women Want in Their Sperm Donor: A Study of More Than 1000 Women's Sperm Donor Selections," *Economics & Human Biology* 23 (2016): 1–9, https://www.ncbi.nlm.nih.gov/pubmed/27359087.

31   Shana Lebowitz, "Science Says Being Tall Could Make You Richer and More Successful—Here's Why," *Business Insider*, September 9, 2015, http://www.businessinsider.com/tall -people-are-richer-and-successful-2015–9; Roger Highfield, "Symmetrical Human Faces Are More Beautiful," *The Telegraph*, June 5, 2008, http://www.telegraph.co.uk/news /science/science-news/3343640/Symmetrical-human-faces-are-more-beautiful.html.

32   R. Sanders et al., "Genome-wide Scan Demonstrates Significant Linkage for Male Sexual Orientation," *Psychological Medicine* 45 (2014): 1379–1388; Siddhartha Mukherjee, *The Gene: An Intimate History* (London: Vintage, 2017): 373–379.

33   Nicholas G. Crawford et al., "Loci Associated with Skin Pigmentation Identified in African Populations," *Science Magazine* 17 (2017), https://www.ncbi.nlm.nih.gov /pubmed/29025994.

34 Kenta Watanabe et al., "CRISPR/Cas9-Mediated Mutagenesis of the *Dihydroflavonol-4-reductase-B (DFR-B)* Locus in the Japanese Morning Glory *Ipomoea (Pharbitis) Nil*," *Scientific Reports* 7 (August 30, 2017), http://bio-engineering.ir/wp-content/uploads/2017/10/s41598-017-10715-1.pdf.

35 "Study Finds Autistics Better at Problem-Solving," EurekAlert!, June 16, 2009, https://www.eurekalert.org/pub_releases/2009–06/uom-sfa061609.php.

36 Neel Burton, "Bipolar Disorder and Creativity," *Psychology Today*, March 9, 2012, https://www.psychologytoday.com/blog/hide-and-seek/201203/bipolar-disorder-and-creativity.

37 Elizabeth Theusch, Analabha Basu, and Jane Gitschier, "Genome-wide Study of Families with Absolute Pitch Reveals Linkage to 8q24.21 and Locus Heterogeneity," *The American Journal of Human Genetics* 85 (2009): 112–119.

38 "Central African Republic," World Food Programme, accessed June 2, 2018, http://www1.wfp.org/countries/central-african-republic; Caroline H. D. Fall, "Fetal Malnutrition and Long-Term Outcomes," in *Maternal and Child Nutrition: The First 1,000 Days*, Nestlé Nutrition Institute Workshop Series, 74 (2013): 11–25, doi:10.1159/000348384.

39 Friedrich Wilhelm Nietzsche, *Thus Spoke Zarathustra* (Chicago: H. Regnery, 1957): 3.

40 Paul Weindling, "Julian Huxley and the Continuity of Eugenics in Twentieth-century Britain," *Journal of Modern European History*, November 1, 2012, accessed May 10, 2018, https://www.ncbi.nlm.nih.gov/pmc/articles/PMC4366572/.

41 Tobiasz Mazan, *Transcend the Flesh: Transhumanism Debate* (2015): 8, https://www.researchgate.net/publication/279189548_Transcend_the_Flesh_Transhumanism_debate.

42 Doug Baily et al., "Transhumanist Declaration," drafted in 1998 and adopted by Humanity+ in 2009, http://humanityplus.org/philosophy/transhumanist-declaration/.

43 Santi Tafarella, "What Did Friedrich Nietzsche Take from Charles Darwin?" *Prometheus Unbound*, 2010, https://santitafarella.wordpress.com/2009/12/26/what-did-friedrich-nietzsche-take-from-charles-darwin/amp/.

CHAPTER 9

1 Shepard Krech III, *The Ecological Indian: Myth and History* (New York: W. W. Norton & Company, 2000). Whether humans are responsible for this extinction, to be fair, is hotly debated among archeologists. For the affirmative case, see Gary Haynes, "The Catastrophic Extinction of North American Mammoths and Mastodonts," *World Archaeology* 33, no. 3 (2002): 391–416, doi:10.1080/0043824012010744 0. For the opposing view, see Donald Grayson, "Clovis Hunting and Large Mammal Extinction: A Critical Review of the Evidence," *Clovis Hunting and Large Mammal Extinction: A Critical Review of the Evidence* 16, no. 4 (December 2012), doi:10.1023/A:1022912030020.

2 1:26 (King James Version), https://www.biblegateway.com/passage/?search=Genesis+1%3A26&version=KJV.

3 Chapter 3, *The Works of Mencius*, http://nothingistic.org/library/mencius/mencius01.html.

4 Frank Dikötter, *Mao's Great Famine: The History of China's Most Devastating Catastrophe, 1958–1962* (New York: Walker & Company, 2010).

5   Elizabeth C. Economy, *The River Runs Black: The Environmental Challenge to China's Future* (Cornell University Press, 2010).

6   Diamond v. Chakrabarty, 447 U.S. 303 (1980), https://supreme.justia.com/cases /federal/us/447/303/case.html.

7   Paul Berg et al., "Potential Biohazards of Recombinant DNA Molecules," *Science* 26 (1974): 303, https://www.ncbi.nlm.nih.gov/pmc/articles/PMC388511/?page=1.

8   Paul Berg et al., "Summary Statement of the Asilomar Conference on Recombinant DNA Molecules," Proceedings of the National Academy of Sciences, 72 (1975): 1981–1984, https://www.ncbi.nlm.nih.gov/pmc/articles/PMC432675/pdf/pnas00049–0007.pdf.

9   Rick Blizzard, "Genetically Altered Foods: Hazard or Harmless?" *Gallup News*, August 12, 2003, http://news.gallup.com/poll/9034/genetically-altered-foods-hazard-harm less.aspx.

10   Cary Funk and Brian Kennedy, "Public Opinion about Genetically Modified Foods and Trust in Scientists Connected with These Foods," Pew Research Center, December 1, 2016, http://www.pewinternet.org/2016/12/01/public-opinion-about-genetically -modified-foods-and-trust-in-scientists-connected-with-these-foods/.

11   "Genetically Modified Seeds Market—9.83% CAGR to 2020," *PR Newswire*, September 7, 2016, https://www.prnewswire.com/news-releases/genetically-modified-seeds -market—-983-cagr-to-2020–592549281.html.

12   For a fantastic overview of the regulations of GM crops in China, see: Alice Yuen-TingWong and Albert Wai-Kit Chan, "Genetically Modified Foods in China and the United States: A Primer of Regulation and Intellectual Property Protection," *Food Science and Human Wellness* 5 no. 3 (2016): 124–140, https://www.sciencedirect.com/science /article/pii/S2213453016300076.

13   Chuin Wei-Yap, "Xi's Remarks on GMO Signal Caution," *Wall Street Journal*, October 9, 2014, https://blogs.wsj.com/chinarealtime/2014/10/09/xis-remarks-on-gmo-signal -caution/.

14   "ChemChina Clinches Its $43 Billion Takeover of Syngenta," *Fortune*, May 5, 2017, http://fortune.com/2017/05/05/chemchina-syngenta-deal-acquisition/.

15   Christina Larson, "Can the Chinese Government Get Its People to Like GMOs?" *New Yorker*, August 31, 2015, https://www.newyorker.com/tech/elements/can-the-chinese -government-get-its-people-to-like-g-m-o-s.

16   Fei Han et al., "Attitudes in China about Crops and Foods Developed by Biotechnology," *Plos One* 10 (2015), http://journals.plos.org/plosone/article?id=10.1371/journal. pone.0139114.

17   "Statement by the AAS Board of Directors on Labeling of Genetically Modified Foods," American Association for the Advancement of Science, June 12, 2013, https://www.aaas .org/news/statement-aaas-board-directors-labeling-genetically-modified-foods.

18   Rod A. Herman and William D. Price, "Unintended Compositional Changes in Genetically Modified (GM) Crops: 20 Years of Research," *Journal of Agricultural and Food Chemistry* 61 (2013): 11695–11701, http://pubs.acs.org/doi/full/10.1021/jf400135r.

19   *A Decade of EU-Funded GMO Research (2001–2010)*, European Commission on Food, Agriculture and Fisheries and Biotechnology, 2010, https://ec.europa.eu/research /biosociety/pdf/a_decade_of_eu-funded_gmo_research.pdf.

20 *Safety Assessment of Foods Derived from Genetically Modified Microorganisms: Report of a Joint FAO/WHO Expert Consultation on Foods Derived from Biotechnology,* WHO Headquarters, Geneva, Switzerland, September 24–28, 2001, http://www.fao.org/3/a -ae585e.pdf.

21 *Report 2 of the Council on Science and Public Health (A-12) Labeling of Bioengineered Foods* (Resolutions 508 and 509-A-11), 2012, http://ag.utah.gov/documents/AMA -BioengineeredFoods.pdf.

22 Read *Safety of Genetically Engineered Foods: Approaches to Assessing Unintended Health Effects* at NAP.edu. National Academies Press: OpenBook, accessed April 30, 2018, https://www.nap.edu/read/10977/chapter/1.

23 *Genetically Modified Plants for Food Use and Human Health—An Update,* Policy Document 4/02, February 2002, https://royalsociety.org/~/media/royal_society_content/policy /publications/2002/9960.pdf.

24 *Genetically Engineered Crops: Experiences and Prospects,* National Academies of Sciences, Engineering, and Medicine (Washington, DC: The National Academies Press, 2016): 15, https://doi.org/10.17226/23395.

25 Elisa Pellegrino et al., "Impact of Genetically Engineered Maize on Agronomic, Environmental and Toxicological Traits: A Meta-analysis of 21 Years of Field Data," *Scientific Reports* 8 (2018), https://www.nature.com/articles/s41598-018-21284-2.

26 Rebecca Goldburg et al., *Biotechnology's Bitter Harvest,* Biotechnology Working Group, March 1990, http://blog.ucsusa.org/wp-content/uploads/2012/05/Biotechnologys -Bitter-Harvest.pdf.

27 For a fair assessment of Monsanto's work, see Drake Bennett, "GMO Factory Monsanto's High-Tech Plans to Feed the World," Bloomberg.com, July 4, 2014, accessed May 10, 2018, https://www.bloomberg.com/news/articles/2014-07-03/gmo-factory-monsantos -high-tech-plans-to-feed-the-world.

28 Eurobarometer Biotechnology Report, October 2010, http://ec.europa.eu /commfrontoffice/publicopinion/archives/ebs/ebs_341_en.pdf.

29 Karl Haro von Mogel, "GMO Crops Vandalized in Oregon," *Biology Fortified,* June 24, 2013, https://www.biofortified.org/2013/06/gmo-crops-vandalized-in-oregon/.

30 Tom Nightingale, "Scientists Speak Out against Vandalism of Genetically Modified Rice," *ABC News,* September 19, 2013, http://www.abc.net.au/news/2013-09-20 /scientists-speak-out-against-vandalism-of-gm-rice/4970626. The fraudulent Indian activist Vandana Shiva also piled onto the anti-GMO bandwagon, claiming that the Green Revolution, which saved up to a billion lives, actually caused hunger and that Golden Rice is a hoax.

31 Mark Lynas, "With G.M.O. Policies, Europe Turns against Science," *New York Times,* October 24, 2015, https://www.nytimes.com/2015/10/25/opinion/sunday /with-gmo-policies-europe-turns-against-science.html?mtrref=www.google. com&assetType=opinion.

32 "EASAC Warns EU Policy on GM Crops Threatens the Future of Our Agriculture," The Royal Netherlands Academy of Arts and Sciences, July 10, 2013, https://www.knaw .nl/en/news/news/easac-warns-eu-policy-on-gm-crops-threatens-the-future-of-our -agriculture.

33 For a damning assessment of the effectiveness of Germany's Green party in undermining

biotechnology innovation in Germany, see Günther, Susanne, "How Anti-GMO Advocates Hijacked German Science, Blocking Agricultural Innovation and Threatening the CRISPR Revolution: A Farmer's Perspective," Genetic Literacy Project, July 17, 2018, accessed August 5, 2018, https://geneticliteracyproject.org/2018/07/12/how-anti -gmo-advocates-hijacked-german-science-blocking-agricultural-innovation-threatening -crispr-revolution-farmers-perspective/.

34   Paul Gander, "EU Enzyme Scrutiny Could Open Up GM 'Can of Worms,'" Food Manufacture, August 8, 2012, https://www.foodmanufacture.co.uk/Article/2012/08/09/ Are-enzymes-processing-aids-or-should-they-be-labelled-as-ingredients.

35   Steven Cerier, "Anti-GMO Forces New Breeding Techniques (NBTs) Despite Similarities to Conventional Crops," Genetic Literacy Project, February 26, 2018, https://geneticliteracyproject.org/2018/02/26/anti-gmo-forces-target-new-breeding -techniques-nbts-despite-similarities-conventional-crops/.

36   "Laureates Letter Supporting Precision Agriculture (GMOs)," Support Precision Agriculture, June 29, 2016, http://supportprecisionagriculture.org/nobel-laureate-gmo -letter_rjr.html.

37   "Organisms Obtained by Mutagenesis Are GMOs and Are, in Principle, Subject to the Obligations Laid down by the GMO Directive." Court of Justice of the European Union, PRESS RELEASE No 111/18, Luxembourg, 25 July 2018, Judgment in Case C-528/16, Confédération paysanne and Others v Premier ministre and Ministre de l'Agriculture, de l'Agroalimentaire et de la Forêt, https://curia.europa.eu/jcms/upload/docs/application /pdf/2018–07/cp180111en.pdf.

38   Callaway, Ewen. "CRISPR Plants Now Subject to Tough GM Laws in European Union," Nature News. July 25, 2018, accessed August 4, 2018, https://www.nature.com/articles /d41586-018-05814-6.

39   Marian Tupy, "Europe's Anti-GMO Stance Is Killing Africans," Reason, September 5, 2017, http://reason.com/archives/2017/09/05/europes-anti-gmo-stance-is-killing-afric.

40   "Laureates Letter Supporting Precision Agriculture (GMOs)," Support Precision Agriculture, June 29, 2016, http://supportprecisionagriculture.org/nobel-laureate -gmo-letter_rjr.html. For more on the dangerous disinformation campaign mounted by anti-GMO organizations against Golden Rice, see Andrew Porterfield, "Anti-GMO Groups Draw FDA Rebuke over Misrepresentation of Golden Rice Nutrition," Genetic Literacy Project, June 15, 2018, accessed June 18, 2018, https://geneticliteracyproject .org/2018/06/11/anti-gmo-groups-draw-fda-rebuke-over-misrepresentation-of-golden -rice-nutrition/?mc_cid=6c723bddda&mc_eid=6d7f502b6d.

41   Researchers at Iowa State University have suggested that news entities backed by the Russian state are pushing anti-GMO news stories in the West. "How Russia Tried to Turn America against GMOs and Agricultural Biotechnology and Sow Ideological Discord," Genetic Literacy Project, March 9, 2018, accessed August 4, 2018, https:// geneticliteracyproject.org/2018/03/06/russia-tried-turn-america-gmos-agricultural -biotechnology-sow-ideological-discord/.

42   National Academies of Sciences. "Science Breakthroughs to Advance Food and Agricultural Research by 2030," National Academies Press: OpenBook, July 18, 2018, accessed August 5, 2018, https://www.nap.edu/catalog/25059/science-breakthroughs-to

-advance-food-and-agricultural-research-by-2030. https://doi.org/10.17226/25059; pp. 4–5.

43   Mireille Jacobson and Heather Royer, "Aftershocks: The Impact of Clinic Violence on Abortion Services," National Bureau of Economic Research, 2010, http://users.nber .org/~jacobson/JacobsonRoyer6.2.10.pdf.

44   "Violence Statistics & History," National Abortion Federation, https://prochoice.org /education-and-advocacy/violence/violence-statistics-and-history/.

45   "Public Funding for Abortion," American Civil Liberties Union, accessed May 1, 2018, https://www.aclu.org/other/public-funding-abortion.

46   Adam Taylor, "The Human Suffering Caused by China's One-Child Policy," *Washington Post*, October 29, 2015, https://www.washingtonpost.com/news/worldviews /wp/2015/10/29/the-human-suffering-caused-by-chinas-one-child-policy/?utm _term=.d7fa61c5c3cb.

47   Justin Parkinson, "Five Numbers That Sum Up China's One-Child Policy," *BBC News*, October 29, 2015, http://www.bbc.com/news/magazine-34666440. The number is probably lower than 400 million because the country's birth rate would almost certainly have decreased in conjunction with economic development and increasing education levels.

48   David Masci, "Where Major Religious Groups Stand on Abortion," Pew Research Center, June 21, 2016, http://www.pewresearch.org/fact-tank/2016/06/21/where-major -religious-groups-stand-on-abortion/.

49   "Views about Abortion," Pew Research Center, 2014, http://www.pewforum.org /religious-landscape-study/views-about-abortion/.

50   Lester Feder, Jeremy Singer-Vine, and Jina Moore, "This Is How 23 Countries around the World Feel about Abortion," *Buzzfeed News*, June 4, 2015, https://www.buzzfeed.com /lesterfeder/this-is-how-23-countries-around-the-world-feel-about-abortio?utm_term=. veJX3nrV4#.swW7R0kQ8. A 2014 Pew poll found that people in the largely Catholic Philippines oppose abortion the most, while residents in mostly secular France oppose it the least. "Global Views on Mortality," Pew Research Center, http://www.pewglobal .org/2014/04/15/global-morality/table/abortion/.

51   Angelina E. Theodorou and Aleksandra Sandstrom, "How Abortion Is Regulated around the World," Pew Research Center, October 6, 2015, http://www.pewresearch.org/fact -tank/2015/10/06/how-abortion-is-regulated-around-the-world/.

52   Gina Pollack, "Undue Burden: Trying to Get an Abortion in Louisiana," *New York Times*, May 16, 2017, https://www.nytimes.com/2017/05/16/opinion/abortion-restrictions -louisiana.html?_r=0.

53   Joshua Seitz, "Striking a Balance: Policy Considerations for Human Germline Modification." *Santa Clara Journal of International Law* 16, no. 1 (March 2, 2018): 60–100, accessed June 18, 2018, https://digitalcommons.law.scu.edu/cgi/viewcontent .cgi?article=1225&context=scujil. For an interesting take on how the gene-editing debate is making strange bedfellows of abortion rights opponents, see Sarah Karlin et al., "Gene Editing: The Next Frontier in America's Abortion Wars," February 16, 2016, accessed May 29, 2018, https://www.politico.com/story/2016/02/gene-editing-abortion-wars -219230.

54   See "Alliance VITA 'Stop GM Babies' Awareness-Raising Campaign Confirmed by the Results of an Opinion Poll on CRISPR-Cas9," Alliance Vita, May 27, 2016, accessed June 24, 2018, https://www.alliancevita.org/en/2016/05/alliance-vita-stop-gm-babies-awareness-raising-campaign-confirmed-by-the-results-of-an-opinion-poll-on-crispr-cas9/.

55   Heather Mason Kiefer, "Gallup Brain: The Birth of In Vitro Fertilization," *Gallup*, August 5, 2003, http://news.gallup.com/poll/8983/gallup-brain-birth-vitro-fertilization.aspx; "Abortion Viewed in Moral Terms: Fewer See Stem Cell Research and IVF as Moral Issues," Pew Research Center, August 5, 2013, http://www.pewforum.org/2013/08/15/abortion-viewed-in-moral-terms/.

56   "Awareness and Knowledge about Reproductive Genetic Technology," Genetics and Public Policy Center, 2002, https://jscholarship.library.jhu.edu/bitstream/handle/1774.2/979/PublicAwarenessAndAttitudes.pdf?sequence=.

57   "The Public and Genetic Editing, Testing, and Therapy," Harvard T. H. Chan School of Public Health, January 2016, https://cdn1.sph.harvard.edu/wp-content/uploads/sites/94/2016/01/STAT-Harvard-Poll-Jan-2016-Genetic-Technology.pdf.

58   Funk, Cary, and Meg Hefferon. "Public Views of Gene Editing for Babies Depend on How It Would Be Used," Pew Research Center: Internet, Science & Tech, July 26, 2018, accessed August 27, 2018, http://www.pewinternet.org/2018/07/26/public-views-of-gene-editing-for-babies-depend-on-how-it-would-be-used/.

59   "The Public and Genetic Editing, Testing, and Therapy."

60   "Results of Major Surveys on Attitudes to Human Genetics," Ipsos Mori, March 1, 2001, https://www.ipsos.com/ipsos-mori/en-uk/results-major-survey-attitudes-human-genetics.

61   "Industry News: UK Public Cautiously Optimistic about Genetic Technologies," *SelectScience*, March 8, 2018, http://www.selectscience.net/industry-news/uk-public-cautiously-optimistic-about-genetic-technologies/?artID=45918. Even though the UK population had come a long way toward accepting human genetic engineering, ironically only 51 percent of those polled supported the use of genetic technologies to increase the efficiency of food production. In other words, a full 25 percent more Britons support genetically modifying their children than do genetically altering a kumquat.

62   Fung-Kei Cheng, "Taijiao: A Traditional Chinese Approach to Enhancing Fetal Growth through Maternal Physical and Mental Health," *Chinese Nursing Research* 3 (2016): 49–53, http://www.sciencedirect.com/science/article/pii/S2095771816300470; Baoqui Su and Darryl R. J. Macer, "Chinese People's Attitudes towards Genetic Diseases and People with Handicaps," *Law and Human Genome Review* 18 (2003): 191–210, http://www.eubios.info/Papers/yousheg.htm.

63   David Cyranoski, "China's Embrace of Embryo Selection Raises Thorny Questions," *Nature* 548 (2017): 272–274, https://www.nature.com/news/china-s-embrace-of-embryo-selection-raises-thorny-questions-1.22468.

64   Jiang-Hui Wang et al., "Public Attitudes toward Gene Therapy in China," *Molecular Therapy Methods & Clinical Development* 6 (2017).

65   Tristan McCaughey et al., "A Global Social Media Survey of Attitudes to Human Genome Editing," *Cell Stem Cell* 18 (2016): 569–572.

66   "Is Pre-implantation Genetic Diagnosis (PGT) Acceptable for Catholics?" Institute

of Catholic Bioethics, January 6, 2014, https://sites.sju.edu/icb/is-pre-implantation
-genetic-diagnosis-PGT-acceptable-for-catholics/.

67   Ioannes Paulus PP. II, "Evangelium Vitae," 1995, http://w2.vatican.va/content/john
-paul-ii/en/encyclicals/documents/hf_jp-ii_enc_25031995_evangelium-vitae.html.

68   Edward Pentin, "The Brave New World of Three-Parent Babies," *National Catholic
Register*, October 2, 2013, http://www.ncregister.com/daily-news/the-brave-new-world
-of-three-parent-babies.

69   "Fearfully and Wonderfully Made: A Policy on Human Biotechnologies," National
Council of Churches, adopted November 8, 2006, http://nationalcouncilofchurches.us
/common-witness/2006/biotech.php.

70   Christopher Benek, "Religion+ for Humanity+," *H+ Magazine*, March 25, 2014, http://
hplusmagazine.com/2014/03/25/religion-for-humanity/.

71   Rabbi Moshe D. Tendler and John D. Loike, "Mitochondrial Replacement Therapy:
Halachic Considerations for Enrolling in an Experimental Clinical Trial," *Rambam
Maimonides Medical Journal* 6 (2015), https://www.ncbi.nlm.nih.gov/pmc/articles
/PMC4524404/.

72   Jaron Lanier, "The First Church of Robotics," *New York Times*, August 9, 2010, http://
www.nytimes.com/2010/08/09/opinion/09lanier.html?_r=1; Jaron Lanier, "Singularity
Is a Religion Just for Digital Geeks," *BigThink*, March 11, 2011, http://bigthink.com
/devils-advocate/singularity-is-a-religion-just-for-digital-geeks.

73   Susannah Baruch, David Kaufman, and Kathy L. Hudson, "Genetic Testing of
Embryos: Practices and Perspectives of U.S. In Vitro Fertilization Clinics," *Fertility and
Sterility* 89 (2008): 1053–1058, https://www.sciencedirect.com/science/article/pii
/S0015028207012162.

74   Bettina Bock von Wülfingen, "Contested Change: How Germany Came to Allow PGT,"
Reproductive Biomedicine and Society Online 3 (2016): 60–67, http://www.sciencedirect
.com/science/article/pii/S2405661816300387.

75   Michelle J. Bayefsky, "Comparative Preimplantation Genetic Diagnosis Policy in Europe
and the USA and Its Implications for Reproductive Tourism," *Reproductive Biomedicine
and Society Online* 3 (2016): 41–47, https://www.sciencedirect.com/science/article/pii
/S2405661817300047.

76   "Cross-Border Reproductive Care: A Committee Opinion," *Fertility and Sterility*
100(2013): 645–650, http://www.fertstert.org/article/S0015–0282(13)00396–8/fulltext.

77   An active debate is ongoing about whether China is a genetic Wild West. For the affirmative
argument, see: Didi Kirsten Tatlow, "A Scientific Ethical Divide Between China and
West," *New York Times*, June 29, 2015, https://www.nytimes.com/2015/06/30/science
/a-scientific-ethical-divide-between-china-and-west.html. For a well-articulated opposing
view, see: Douglas Sipp and Duanqing Pei, "Bioethics in China: No Wild East," *Nature*
534 (2016): 465–467, https://www.nature.com/news/bioethics-in-china-no-wild
-east-1.20116; and Ian Johnson and Cao Li, "China Experiences a Booming Underground
Market in Surrogate Motherhood," *New York Times*, August 2, 2014, https://www
.nytimes.com/2014/08/03/world/asia/china-experiences-a-booming-black-market
-in-child-surrogacy.html. See also Genome Editing and Human Reproduction.
Report. July 17, 2018, Nuffield Council on Bioethics. http://nuffieldbioethics.org/wp

-content/uploads/Genome-editing-and-human-reproduction-FINAL-website.pdf, 110–111.

78 Alta R. Charo, "On the Road (to a Cure?)—Stem-Cell Tourism and Lessons for Gene Editing," *New England Journal of Medicine* 374 (2016): 901–903, http://www.nejm.org /doi/full/10.1056/NEJMp1600891.

79 Mahsa Shabani and Pascal Borry, "Rules for Processing Genetic Data for Research Purposes in View of the New EU General Data Protection Regulation," *European Journal of Human Genetics* 26, no. 2 (2017): 149–56, doi:10.1038/s41431-017-0045-7.

80 Richard Bird, "Where Are We Now with Data Protection Law in China?" Freshfields Bruckhaus Deringer, 2017, https://www.freshfields.com/en-us/our-thinking /campaigns/digital/data/where-are-we-now-with-data-protection-law-in-china/.

81 Wenxin Fan, Natasha Khan, and Liza Lin, "China Snares Innocent and Guilty Alike to Build World's Biggest DNA Database," *Wall Street Journal*, December 26, 2017, https:// www.wsj.com/articles/china-snares-innocent-and-guilty-alike-to-build-worlds-biggest -dna-database-1514310353; Echo Huang, "China Is Creating a Massive 'Orwellian' DNA Database," *Quartz*, May 16, 2017, https://qz.com/984400/china-is-creating-a-massive -orwellian-dna-database-to-construct-harmonic-society/; "China: Police DNA Database Threatens Privacy," May 15, 2017, Human Rights Watch, https://www.hrw.org /news/2017/05/15/china-police-dna-database-threatens-privacy. For more on repression in Xinjiang, including the Uighur population being required to submit blood samples, see "China Has Turned Xinjiang into a Police State Like No Other," *The Economist*, May 31, 2018, accessed June 3, 2018, https://www.economist .com/briefing/2018/05/31/china-has-turned-xinjiang-into-a-police-state-like-no -other?frsc=dg|e. In August 2019, a UN Human Rights panel announced it had received credible reports that China was holding approximately one million Uighurs in "massive internment camps." Reuters. "U.N. Says It Has Credible Reports That China Holds Million Uighurs in Secret Camps." *New York Times*, August 10, 2018, accessed August 15, 2018, https://www.nytimes.com/reuters/2018/08/10/world/asia/10reuters-china-rights-un .html. Not surprisingly, Beijing quickly refuted this claim. "China Has Prevented 'Great Tragedy' in Xinjiang, State-Run Paper Says," *Reuters*, August 13, 2018, accessed August 15, 2018, https://www.reuters.com/article/us-china-rights-un/china-has-prevented -great-tragedy-in-xinjiang-state-run-paper-says-idUSKBN1KY01B?il=0. The UK Home Office was in 2018 also exploring the possibility of creating a centralized database of all the biometric data collected from UK citizens.

CHAPTER 10

1   Harvard psychologist Steven Pinker makes a compelling case for this point. Steven Pinker, *Enlightenment Now: The Case for Reason, Science, Humanism, and Progress* (London: Allen Lane, 2018).

2   Because ill-informed concepts of genetics have underpinned racism, colonialism, eugenics, and discrimination for centuries, discussing genetics and particular outcomes like sports prowess is rightly sensitive. For thoughtful considerations of how tricky it can be to explore genetics and race see: David Reich, "How Genetics Is Changing Our

Understanding of 'Race,'" *New York Times*, March 23, 2017, https://www.nytimes.com/2018/03/23/opinion/sunday/genetics-race.html and Siddhartha Mukherjee, *The Gene: An Intimate History* (London: Vintage, 2017), 341–345.

3   David J. Epstein, *The Sports Gene: What Makes the Perfect Athlete* (London: Yellow Jersey Press, 2013). See also: A. de la Chapelle, A. L. Traskelin, and E. Juvonen, "Truncated Erythropoietin Receptor Causes Dominantly Inherited Benign Human Erythrocytosis," *Proceedings of the National Academy of Sciences* 90 (1993): 4495–4499, https://www.ncbi.nlm.nih.gov/pmc/articles/PMC46538/pdf/pnas01462-0175.pdf; David Epstein, "Eero Mäntyranta–Finland's Champion. 1937–2013: Obituary," *The Science of Sport*, December 31, 2013, http://sportsscientists.com/2013/12/eero-mantyranta-finlands-champion-1937-2013-obituary/.

4   Eynon et al., "Genes for Elite Power and Sprint Performance: ACTN3 Leads the Way." *Sports Medicine*, September 2013, accessed May 1, 2018, https://www.ncbi.nlm.nih.gov/pubmed/23681449.

5   Ross Tucker, Vincent O. Onywera, and Jordan Santos-Concejero, "Analysis of the Kenyan Distance-Running Phenomenon," *International Journal of Sports Physiology and Performance* 10 (2015): 285–291, https://www.researchgate.net/publication/264745551_Analysis_of_the_Kenyan_Distance-Running_Phenomenon.

6   To be clear, this success was not based on genetics alone but instead resulted from the complex interaction of genotypic, phenotypic, and socioeconomic factors. This topic is hotly debated. See Max Fisher, "Why Kenyans Make Such Great Runners: A Story of Genes and Cultures." *The Atlantic*, April 17, 2012, accessed June 6, 2018, https://www.theatlantic.com/international/archive/2012/04/why-kenyans-make-such-great-runners-a-story-of-genes-and-cultures/256015/; and Alex Hutchinson, "Kenyan Dominance, Real and Imagined," *Runner's World*, May 25, 2018, accessed June 6, 2018, https://www.runnersworld.com/training/a20846404/kenyan-dominance-real-and-imagined/.

7   Andrew Roos and Thomas Roos, "Genetics of Athletic Performance," Stanford University, May 15, 2012, https://web.stanford.edu/class/gene210/files/projects/Gene210-Athletics Presentation-Roos.pdf.

8   Olympic officials for the first time tested athletes for "gene doping" at the 2016 Rio Olympics to see if they had evidence of gene therapies being used to increase a given athlete's ability to produce red blood cells. Eric Niiler, "Olympic Drug Cops Will Scan For Genetically Modified Athletes," *Wired*, July 28, 2016, https://www.wired.com/2016/07/olympic-drug-cops-will-scan-genetically-modified-athletes/.

9   Nick Webborn et al., "Direct-to-Consumer Genetic Testing for Predicting Sports Performance and Talent Identification: Consensus Statement," *British Journal of Sports Medicine* 49 (2015): 1486–1491, https://www.ncbi.nlm.nih.gov/pubmed/26582191.

10   Ron Synovitz and Zamira Eshanova, "Uzbekistan Is Using Genetic Testing to Find Future Olympians," *The Atlantic*, February 6, 2014, https://www.theatlantic.com/international/archive/2014/02/uzbekistan-is-using-genetic-testing-to-find-future-olympians/283001/.

11   Hoon Choi and Lvaro Choi, "When One Door Closes: The Impact of the Hagwon Curfew on the Consumption of Private Tutoring in the Republic of Korea,"

SSRN Electronic Journal (2015), http://www.ub.edu/irea/working_papers/2015
/201526.pdf.

12   Malleta, King, "Why Korean Parents Give Their Kids Plastic Surgery as Graduation
Gifts," NextShark, December 23, 2016, accessed August 9, 2018, https://nextshark.
com/why-korean-parents-give-their-kids-plastic-surgery-as-graduation-gifts/; Wei, Will,
"Why Korean Parents Are Having Their Kids Get Plastic Surgery before College." *Business
Insider*, November 25, 2017, accessed August 09, 2018, https://www.businessinsider.
com/eyelid-surgery-in-south-korea-2015–11.

13   Patricia Marx, "About Face," *New Yorker*, March 23, 2015, https://www.newyorker.com
/magazine/2015/03/23/about-face.

14   William Wan, "In China, Parents Bribe to Get Students into Top Schools, Despite
Campaign against Corruption," *Washington Post*, October 7, 2013, https://www
.washingtonpost.com/world/in-china-parents-bribe-to-get-students-into-top-schools
-despite-campaign-against-corruption/2013/10/07/fa8d9d32–2a61–11e3–8ade
-a1f23cda135e_story.html?utm_term=.c79e7e712e12.

15   Kitty Bu and Maxim Duncan, "Playtime a Luxury for Competitive Chinese Kids," *Reuters*,
November 23, 2009, https://www.reuters.com/article/us-china-children-play/playtime
-a-luxury-for-competitive-chinese-kids-idUSTRE5AM16920091123.

16   Peter Foster, "Third of Chinese Primary School Children Suffer Stress, Study Finds,"
The Telegraph, January 19, 2010, https://www.telegraph.co.uk/news/worldnews/asia
/china/7027377/Third-of-Chinese-primary-school-children-suffer-stress-study-finds.html.

17   Alvin A. Rosenfeld and Nicole Wise, *The Over-Scheduled Child: Avoiding the Hyper-
Parenting Trap* (New York: St. Martins Griffin, 2001).

18   Timothy Caulfield et al., "Marginally Scientific? Genetic Testing of Children and
Adolescents for Lifestyle and Health Promotion," *Journal of Law and the Biosciences* 2
(2015): 627–644, https://www.ncbi.nlm.nih.gov/pmc/articles/PMC5034400/.

19   Antonio Regalado, "Baby Genome Sequencing for Sale in China," *MIT Technology
Review*, June 15, 2017, https://www.technologyreview.com/s/608086/baby-genome
-sequencing-for-sale-in-china/.

20   Kalokairinou et al., "Legislation of Direct-to-Consumer Genetic Testing in Europe: A
Fragmented Regulatory Landscape," *Journal of Community Genetics* 9 (2018): 117–132.
In the United States, companies like Salesforce, OpenTable, and Snap now offer their
employees free genetic screening to assess hereditary disease risk. In 2018, creative
investigators set up fake consumer genetic profiles based on DNA samples found at crime
scenes four decades ago to identify the genetic relatives of and ultimately catch the Golden
State Killer. Clearly, consumer genetic information clearly cannot be a free-for-all. Megan
Molteni, "The Creepy Genetics Behind the Golden State Killer Case," *Wired*, May 4, 2018,
accessed May 29, 2018, https://www.wired.com/story/detectives-cracked-the-golden-
state-killer-case-using-genetics/.

21   S. Roberts et al., "Direct-to-Consumer Genetic Testing: User Motivations, Decision
Making, and Perceived Utility of Results," Public Health Genomics 20, no. 1, (January 10,
2017): 36–45, accessed May 1, 2018. https://www.ncbi.nlm.nih.gov/pubmed/28068660.
For a full list of references on this topic, see http://www.genomes2people.org/wp
-content/uploads/2018/01/PGen_Publications.pdf.

22 Francis Fukuyama, "The End of History?" *The National Interest*, no. 16 (Summer 1989): 3–18, http://www.jstor.org/stable/24027184.

23 Elsa B. Kania, *Battlefield Singularity: Artificial Intelligence, Military Revolution, and China's Future Military Power*, Center for a New American Security, November 2017, 51, https://s3.amazonaws.com/files.cnas.org/documents/Battlefield-Singularity-November-2017.pdf?mtime=20171129235804.

24 *The National Artificial Intelligence Research and Development Strategic Plan*, National Science and Technology Council, October 2016, https://obamawhitehouse.archives.gov/sites/default/files/whitehouse_files/microsites/ostp/NSTC/national_ai_rd_strategic_plan.pdf.

25 Issued by the State Council New Generation Artificial Intelligence Development Planning Notice No. 35 [2017] The People's Governments of Provinces, Autonomous Regions, and Municipalities Directly under the Central Government, the Ministries and Commissions of the State Council, and the Agencies Directly under the State Council, July 8, 2017, accessed June 18, 2018, http://www.gov.cn/zhengce/content/2017–07/20/content_5211996.htm.

26 Ma Si, "Key AI Guidelines Unveiled," *China Daily*, December 15, 2017, http://www.chinadaily.com.cn/a/201712/15/WS5a330a41a3108bc8c6734c64.html; Neil Connor, "Anxious Chinese Parents Cause Gene Testing Boom as They Try to Discover Young Children's Talents," *The Telegraph*, February 11, 2017, http://www.telegraph.co.uk/news/2017/02/11/anxious-chinese-parents-fuel-gene-testing-boom-try-discover/.

27 "Chinese AI Startups Scored More Funding Than America's Last Year," *MIT Technology Review*, February 14, 2018, https://www.technologyreview.com/the-download/610271/chinas-ai-startups-scored-more-funding-than-americas-last-year/.

28 Graham Webster et al., "China's Plan to 'Lead' in AI: Purpose, Prospects, and Problems," New America, August 1, 2017, https://www.newamerica.org/cybersecurity-initiative/blog/chinas-plan-lead-ai-purpose-prospects-and-problems/.

29 See Coral Davenport, "In the Trump Administration, Science Is Unwelcome. So Is Advice," *New York Times*, June 9, 2018, accessed June 18, 2018, https://www.nytimes.com/2018/06/09/climate/trump-administration-science.html.

30 Chappellet-Lanier, Tajha. "White House Announces Creation of Select Committee on Artificial Intelligence," Fedscoop, May 10, 2018, accessed August 09, 2018, https://www.fedscoop.com/white-house-artifical-intelligence-committee-kratsios/.

31 David Cyranoski, "China's Bid to Be a DNA Superpower," *Nature* 534 (2016): 462–463, http://www.nature.com/news/china-s-bid-to-be-a-dna-superpower-1.20121.

32 Susan Decker, "China Becomes One of the Top 5 U.S. Patent Recipients for the First Time," *Bloomberg News*, January 9, 2018, https://www.bloomberg.com/news/articles/2018-01-09/china-enters-top-5-of-u-s-patent-recipients-for-the-first-time.

33 Reinhilde Veugelers, "China Is the World's New Science and Technology Powerhouse," *Bruegel*, August 30, 2017, http://bruegel.org/2017/08/china-is-the-worlds-new-science-and-technology-powerhouse/. Ben Guarino, Emily Rauhala, and William Wan, "China Increasingly Challenges American Dominance of Science," *Washington Post*, June 3, 2018, accessed June 6, 2018, https://www.washingtonpost.com/national/health-science/china-challenges-american-dominance-of-science/2018/06/03/c1e0cfe4–48d5–11e8–827e-190efaf1f1ee_story.html.

34   Kai-Fu Lee, "China's Artificial Intelligence Revolution: Understanding Beijing's Structural Advantages," Sinovation Ventures (December 2017), https://www.eurasiagroup.net/files/upload/China_Embraces_AI.pdf.

35   Eleonore Pauwels and Apratim Vidyarthi, "Who Will Own the Secrets in Our Genes? A U.S.-China Race in Artificial Intelligence and Genomics," Wilson Center, February 2017, https://www.wilsoncenter.org/sites/default/files/who_will_own_the_secrets_in_our_genes.pdf.

36   For a healthy debate on this topic see: Jennifer Kulynych and Henry T. Greely, "Clinical Genomics, Big Data, and Electronic Medical Records: Reconciling Patient Rights with Research When Privacy and Science Collide," *Journal of Law and the Biosciences* 4 (2017): 94–132, https://www.ncbi.nlm.nih.gov/pubmed/28852559. See also: Eryn Brown, "Geneticist on DNA Privacy: Make It So People Don't Care," *Los Angeles Times*, January 18, 2013, http://articles.latimes.com/2013/jan/18/science/la-sci-sn-george-church-dna-genome-privacy-20130118. Innovative American companies might also find a way of using blockchain and other technologies to stitch together disparate data pools while better protecting individual privacy or of convincing large numbers of informed citizens to opt in.

37   iCarbonX, https://www.icarbonx.com/.

38   David Ewing Duncan, "Can AI Keep You Healthy?" *MIT Technology Review*, October 3, 2017, https://www.technologyreview.com/s/608987/how-ai-will-keep-you-healthy/.

39   Jun Wang, "How Digital DNA Could Help You Make Better Health Choices," TED17, https://www.ted.com/talks/jun_wang_how_digital_dna_could_help_you_make_better_health_choices/transcript#t-150728.

40   Jun Wang, "The Bank of Life Gene Concerning Everyone Is Now Coming," *Medium*, April 25, 2016, https://medium.com/@iAskMedia/wang-jun-the-bank-of-life-gene-concerning-everyone-is-now-coming-335353aaf3ee.

41   Henny Sender, "China's Tech Groups Are Building Too Much Power," *Financial Times*, August 28, 2017, https://www.ft.com/content/858d0312-8988-11e7–8bb1–5ba57d47eff7.

42   Tamar Lewin, "Coming to U.S. for Baby, and Womb to Carry It," *New York Times*, July 5, 2014, https://www.nytimes.com/2014/07/06/us/foreign-couples-heading-to-america-for-surrogate-pregnancies.html?hpw&action=click&pgtype=Homepage&version=Hp HedThumbWell&module=well-region&region=bottom-well&WT.nav=bottom-well.

43   "List of Invasions," *Wikipedia*, last modified July 28, 2018, https://en.wikipedia.org/wiki/List_of_invasions.

44   Robert Farley et al., "China vs. America: 3 Ways a War in the South China Sea Could Start," *The National Interest*, accessed June 24, 2018, http://nationalinterest.org/blog/the-buzz/china-vs-america-3-ways-war-the-south-china-sea-could-start-26034.

CHAPTER 11

1   Rob Carlson, "Splice it Yourself," *Wired*, May 1, 2005, https://www.wired.com/2005/05/splice-it-yourself/.

2   Bart Kolodziejczyk, "Do-It-Yourself Biology Shows Safety Risks of an Open Innovation Movement," *Brookings Institute Techtank Blog*, October 9, 2017, https://www.brookings.edu/blog/techtank/2017/10/09/do-it-yourself-biology-shows-safety-risks-of-an-open-innovation-movement/. See also: Mallory Locklear, "These Kids Are Learning CRISPR

At Summer Camp," *Motherboard*, July 27, 2017, https://motherboard.vice.com/en_us/article/kzavja/these-kids-are-learning-crispr-at-summer-camp. Additional information on this topic was provided by Ray McCauley in an email to the author, November 13, 2018.

3   Lisa C. Ikemoto, "DIY Bio: Hacking Life in Bioetech's Backyard," *UC Davis Law Review* 51 (2017): 539–568, https://lawreview.law.ucdavis.edu/issues/51/2/Symposium/51-2_Ikemoto.pdf.

4   See Emily Baumgaertner, "As D.I.Y. Gene Editing Gains Popularity, 'Someone Is Going to Get Hurt,'" *New York Times*, May 14, 2018, accessed May 31, 2018, https://www.nytimes.com/2018/05/14/science/biohackers-gene-editing-virus.html.

5   Siddhartha Mukherjee, *The Gene: An Intimate History* (London: Vintage, 2017), 467.

6   Kate Charlet, "The New Killer Pathogens: Countering the Coming Bioweapons Threat," *Foreign Affairs*, May/June 2018, https://www.foreignaffairs.com/articles/2018-04-16/new-killer-pathogens; Akshat Rathi, "This Could Be the Next Weapon of Mass Destruction," *Quartz*, November 20, 2015, accessed May 3, 2018. https://qz.com/554337/this-could-be-the-next-weapon-of-mass-destruction/.

7   Ryan S. Noyce, Seth Lederman, and David H. Evans, "Construction of an Infectious Horsepox Virus Vaccine from Chemically Synthesized DNA Fragments," *Plos One* 13, no. 1 (2018), doi:10.1371/journal.pone.0188453.

8   Sharon Begley, "Why FBI and the Pentagon Are Afraid of Gene Drives," STAT, April 19, 2018, accessed May 11, 2018, https://www.statnews.com/2015/11/12/gene-drive-bioterror-risk/.

9   James R. Clapper, "Worldwide Threat Assessment of the U.S. Intelligence Community," Senate Armed Services Community (February 9, 2016), 9, https://www.dni.gov/files/documents/SASC_Unclassified_2016_ATA_SFR_FINAL.pdf.

10   "Global Trends: Paradox of Progress," National Intelligence Council (January 2017), https://www.dni.gov/files/documents/nic/GT-Full-Report.pdf.

11   National Academies of Sciences, Engineering, and Medicine. *Biodefense in the Age of Synthetic Biology* (Washington, DC: National Academies Press, 2018). https://www.nap.edu/catalog/24890/biodefense-in-the-age-of-synthetic-biology, doi:10.17226/24890.

12   Richard A. Clarke and R. P. Eddy, *Warnings: Finding Cassandras to Stop Catastrophes* (Harper Collins, 2017).

13   Amy Gutmann and Jonathan Moreno, "Keep CRISPR Safe: Regulating a Genetic Revolution," *Foreign Affairs*, April 25, 2018, accessed May 3, 2018, https://www.foreignaffairs.com/articles/2018-04-16/keep-crispr-safe.

14   Some of the more important meetings have been held in Napa, California, Manchester, England, Washington, DC, and Hong Kong. "International Summit on Human Gene Editing: A Global Discussion," *National Academies Press*, December 1–3, 2015, https://www.ncbi.nlm.nih.gov/books/NBK343651/; *Human Genome Editing: Science, Ethics, and Governance* (Washington, DC: National Academies Press, 2017): 5, 8–9; "Statement on Genome Editing Technologies and Human Germline Genetic Modification," The Hinxton Group, Sept 3–4, 2015, http://www.hinxtongroup.org/Hinxton2015_Statement.pdf.

15   Cited in *Human Enhancement*, ed. Nick Bostrom and Julian Savulescu (Oxford: Oxford University Press, 2009): 132.

16   Jamie Metzl, "Brave New World War," *Democracy* 8, no. 8 (Spring 2008), https://
     democracyjournal.org/magazine/8/brave-new-world-war/.

17   "UNESCO Panel of Experts Calls for Ban on 'Editing' of Human DNA to Avoid Unethical
     Tampering with Hereditary Traits," UNESCO, 2015, https://en.unesco.org/news/unes
     co-panel-experts-calls-ban-editing-human-dna-avoid-unethical-tampering-hereditary-traits.

18   "Statement on Genome Editing Technologies," Committee on Bioethics (2015),
     document DH-BIO/INF (2015) 13, https://rm.coe.int/168049034a.

19   The United Nations Office on Genocide Prevention and the Responsibility to Protect,
     http://www.un.org/en/genocideprevention/about-responsibility-to-protect.html.

20   Edward O. Wilson, *The Social Conquest of Earth* (New York: Liveright Publishing,
     2013): 7.

21   Genome Editing—Progress Educational Trust, accessed August 5, 2018, https://www
     .progress.org.uk/genomeediting.

22   Genome Editing and Human Reproduction, Report, July 17, 2018, Nuffield Council
     on Bioethics, http://nuffieldbioethics.org/wp-content/uploads/Genome-editing-and
     -human-reproduction-FINAL-website.pdf, 142.

Additional Reading

Some excellent books explore many of the individual topics I discuss in *Hacking Darwin*. A few of my favorites include:

Better Than Human by Allen Buchanan, *Regenesis* by George Church and Ed Regis, *A Crack in Creation* by Jennifer Doudna, *Evolving Ourselves* by Juan Enriquez and Steve Gullans, *Radical Evolution* by Joel Garreau, *The End of Sex and the Future of Reproduction* by Hank Greely, *Homo Deus* by Yuval Noah Harari, *How We Do It* by Robert Martin, *The Gene* by Siddhartha Mukherjee, *Blueprint* by Robert Plomin, *Who We Are and How We Got Here* by David Reich, and *She Has Her Mother's Laugh* by Carl Zimmer.

Because the genetic revolution is unfolding so quickly, there are many incredible (and faster-moving) websites, blogs, and podcasts that are essential resources very much worth exploring.

Index

Acknowledgments

I could not have written this book without the help of some truly incredible people to whom I am eternally grateful. Special thank-yous to my brilliant research assistant, Nicola Morrow, for pulling together critical background materials as well as to Yujia He for helping track down important information about developments in China. Nir Barzilai, Serena Chen, George Church, Robert Green, Houman Hemmati, Stephen Hsu, Matt Kaeberlein, Jay Menitove, David Sable, Nathan Treff, and Rakhi Varma all read earlier versions of the manuscript and provided extremely useful comments. My super-efficient and capable agent, Jill Marsal, helped me find the perfect editor and publisher for the book and channeled her inner Sigmund Freud when I first sat down in front of a blank screen and got a little nervous. I can't say enough about my phenomenal editor at Sourcebooks, Grace Menary-Winefield. Grace is among the best editors I've ever encountered. Her passion for this book and for the art of editing and publishing more generally played a critical role in helping my idea reach its potential. Liz Kelsch, Lizzie Lewandowski, and Cassie Gutman, as well as the rest of the team at Sourcebooks, also did incredible work bringing *Hacking Darwin* to life and out into the world. Thank you also to the many thousands of people who attended my talks over recent years on the subjects covered in the book and who challenged me with questions and comments that expanded my thinking. I dedicate the book to the loving memory of Scott Newman, the loving memory of Irwin Blitt, to my parents, Kurt and Marilyn Metzl, to my wonderful nieces Anna Rose and Clara Bea Metzl, and to Mallika Bhargava.

About the Author

Jamie Metzl is a technology futurist and geopolitical expert, media commentator, and senior fellow of the Atlantic Council. He previously served in the U.S. National Security Council, the State Department, the Senate Foreign Relations Committee, as a human rights officer for the United Nations in Cambodia, as executive vice president of the Asia Society, and as chief strategy officer for a biotechnology company. Jamie appears regularly on national and international media, and his syndicated columns and other writing on international affairs, genetics, virtual reality, and other topics are featured regularly in publications around the world. In addition to *Hacking Darwin*, he is the author of a history of the Cambodian genocide, the historical novel *The Depths of the Sea*, and the genetics sci-fi thrillers *Genesis Code* and *Eternal Sonata*.

A member of the Council on Foreign Relations and a former White House fellow and Aspen Institute Crown fellow, Jamie holds a PhD from Oxford, a JD from Harvard Law School, and is a magna cum laude Phi Beta Kappa graduate of Brown University. He lives in New York City and is an avid Ironman triathlete and ultramarathoner.

Visit Jamie online at jamiemetzl.com.